Lecture Notes in Mathematics

A collection of informal reports and seminars

Edited by A. Dold, Heidelberg and B Eckmann, Zürich

311

Conference on Commutative Algebra

Lawrence, Kansas 1972

Edited by James W. Brewer and Edgar A. Rutter

The University of Kansas, Lawrence, KS/USA

Springer-Verlag

Berlin · Heidelberg · New York 1973

AMS Subject Classifications (1970): 13-02, 13A05, 13B20, 13Cxx, 13D05, 13E05, 13E99, 13F05, 13F20, 13G05, 13H10, 13H15, 13H99, 13J05

ISBN 3-540-06140-1 Springer-Verlag Berlin · Heidelberg · New York
ISBN 0-387-06140-1 Springer-Verlag New York · Heidelberg · Berlin

Offsetdruck: Julius Beltz, Hemsbach/Bergstr.

PREFACE

This volume contains a collection of articles contributed to a Conference on Commutative Algebra held at The University of Kansas during the week of May 8 - May 12 1972. The Conference received funding from the National Science Foundation under Grant No. G.P. - 33192 and we would like to express our appreciation to the Foundation for its support.

To make possible the early appearance of these Proceedings, and thereby to insure their timeliness, the volume has been prepared directly from the typewritten manuscripts supplied by the contributors. In instances where the correction of minor errors would have led to a delay in the appearance of the Proceedings, the editors chose to include the paper as received rather than delay its appearance by returning it to the author for correction. We therefore take the responsibility for all misprints which may bedevil the reader.

We should like to thank each of the Conference participants for their enthusiasm and cooperation. Special thanks must go to Professors Shreeram Abhyankar, David Buchsbaum, Robert Gilmer, Irving Kaplansky and Pierre Samuel who were involved in the Conference from its outset and without whose assistance it would not have been possible. Special thanks also must go to our Kansas colleagues, Paul Conrad, Phil Montgomery and Paul McCarthy for their advice and help. Finally we wish to thank Deborah Phelps for her efficiency in attending to the countless details involved in such an undertaking.

James W. Brewer
Edgar A. Rutter

Lawrence, August, 1972

TABLE OF CONTENTS

PARTICIPANTS

Abhyankar, S.
Albis, V.
Arnold, J. T.
Barger, S. F.
Bennett, B.
Bertholf, D.
Bittman, R.
Blass, J.
Boisen, M.
Boorman, E.
Brown, D.
Buchsbaum, D.
Butts, H. S.
Davis, E. D.
Eagon, J.
Eakin, P.
Eggert, N.
Eisenbud, D.
Enochs, E.
Entyne, S.
Epp, S.
Evans, G.
Fields, D.
Fischer, K.
Gabel, M.
Geramita, A.
Gilbert, J.
Gilmer, R.
Gover, E.
Grams, A.
Greenberg, B.
Hays, J.
Hedstrom, J.
Heinzer, W.
Heitman, R.
Hinkle, G.
Hochster, M.
Huckaba, J.
Kaplansky, I.
Kelly, P.
Kohn, P.
Kuan, Wei-Eihn
Larsen, M.
Larson, L.
Levin, G.
Lewis, J.
Lu, Chin-Pi
MacAdam, S.
McDonald, B.
MacRae, R.
Magarian, E.
Mott, J.
Murthy, M.
Nichols, J.
Ohm, J.
O'Malley, M.
Orbach, K.
Orzech, G.
Orzech, M.
Pacholke, K.
Parker, T.
Prekowitz, B.
Ratliff, L.
Razar, M.
Robson, J. C.
Sally, J.
Samuel, P.
Sathaye, A.
Shannon, D.
Sheets, R.
Sheldon, P.
Shores, T.
Smith, W.
Snider, R.
Stromquist, W.
Thaler, A.
Wadsworth, A.
Wichman, M.
Wiegand, R.
Wiegand, S.
Woodruff, D.
Van der Put, M.
Vasconcelos, W.
Vekovius, A.

ON MACAULAY'S EXAMPLES

Lectures by Professor Abhyankar

Notes by A. Sathaye

§1 INTRODUCTION AND PRELIMINARIES

Introduction. In [1] p. 36 Macaulay has given examples of prime ideals P_m in the polynomial ring in 3 variables which need at least m generators. In fact all the examples are ideals defining irreducible curves in affine 3-space with singularities at the origin. The same ideals extended to the local ring of the origin in affine 3-space will still need at least m generators, thus providing examples in a regular local ring of dimension 3.

It seems however that the proof, although precise, is written with many assertions which were probably part of the 'well known' in Macaulay's time, but are rather mysterious today and seem in need of proof. After rediscovering the proofs and reexplaining them to various individuals over the years, Professor Abhyankar thought it would be a worthwhile public service to publish an explanatory note.

The note was prepared by A. Sathaye from the Purdue Seminar by and discussions with Professor Abhyankar. Also P. Russell offered many suggestions in the writing of the note.

In the remaining part of §1 we fix up terminology and prove Proposition 1 — a residuation theorem to be used later on.

In §2 we construct a set of $\frac{1}{2}m(m-1)$ points in the plane (in fact on a certain curve f) so that it does not lie on any curve of degree $(m-2)$ and the curves of degree $(m-1)$ through it do not have any fixed point outside. This construction of the set, besides being useful for the construction of the examples, raises some questions which are collected in §4.

§3 provides the detailed proof of the following theorem along the same lines as in Macaulay's original proof.

Theorem: Let k be an algebraically closed field and $A = k[X, Y, Z]$, the polynomial ring in 3 variables over k. Then for each $m \geq 1$ there exists a prime ideal $P_m \subset k[X, Y, Z]$ so that P_m needs at least m generators.

Note: We note that for $m = 1, 2$ one can take the ideal of an irreducible surface and an irreducible curve respectively and hence we assume $m \geq 3$. After fixing m we shall denote the ideal P_m by P.

1.1 General Terminology

Let k be an algebraically closed field. Let $R^{(n)} = k[X_0, \ldots, X_n]$ be the graded polynomial ring in $(n + 1)$ variables over k. Then $R^{(n)} = \sum_{i=0}^{\infty} R_i^{(n)}$ where $R_i^{(n)} = \{u \mid u$ homogeneous of degree $i\}$ where 0 is homogeneous of all degrees.

Let $H \in R_i^{(n)}$ for some i so that $H \neq 0$ and H is irreducible. Denote the images of various quantities in $R^{(n)}$ modulo H by a bar $(\overline{\ })$. $\overline{R^{(n)}}$ is the homogeneous coordinate ring of the hypersurface H of degree i in projective n-space. Let $\Omega(H)$ denote $\{\frac{u}{v} \mid u, v \in \overline{R^{(n)}},\ u, v$ homogeneous of the same degree, $v \neq 0\}$, the function field of H.

Let $\mathfrak{m} \subset \overline{R^{(n)}}$ be a homogeneous minimal prime ideal. The local ring of H at $\mathfrak{m}$ is defined by $\mathcal{O}_{\mathfrak{m}} = \{\frac{u}{v} \in \Omega(H) \mid v \notin \mathfrak{m}\}$.

We shall consider only hypersurfaces H which are nonsingular in codimension 1, i.e. those for which for all $\mathfrak{m}$ as above $\mathcal{O}_{\mathfrak{m}}$ is a prime divisor of $\Omega(H)$ and then in fact a prime divisor of the first kind with respect to H. Thus there is a one to one correspondence between minimal homogeneous prime ideals in $\overline{R^{(n)}}$ and prime divisors of $\Omega(H)$ of the first kind with respect to H given by $\mathfrak{m} \longleftrightarrow \mathcal{O}_{\mathfrak{m}}$. Thus we may interchange $\mathfrak{m}$ and $\mathcal{O}_{\mathfrak{m}}$ depending on the context. Let $\mathfrak{S}(H)$ denote the set of all minimal homogeneous prime ideals in $\overline{R^{(n)}}$. A divisor D on H is a map $D : \mathfrak{S}(H) \longrightarrow Z$, the group of integers so that $D(\mathfrak{m}) = 0$ for all but finitely many $\mathfrak{m} \in \mathfrak{S}(H)$. The divisors clearly form a group by defining $(D + E)(\mathfrak{m}) = D(\mathfrak{m}) + E(\mathfrak{m})$. We denote the group of all divisors by $\mathfrak{D}(H)$. Define $D \geq E$ if and only if $D(\mathfrak{m}) \geq E(\mathfrak{m})$ for all $\mathfrak{m}$. Then the semigroup of divisors D so that $D \geq 0$ is denoted by $\mathfrak{P}(H)$ and divisors in $\mathfrak{P}(H)$ are called effective.

For $\mathfrak{m} \in \mathfrak{S}(H)$ we shall denote by $v_{\mathfrak{m}}$ the valuation associated with $\mathcal{O}_{\mathfrak{m}}$. For a set of homogeneous elements $S \subset \overline{R^{(n)}}$ define $S\mathcal{O}_{\mathfrak{m}} = \{\frac{u}{v} \mid u \in S,\ \frac{u}{v} \in \mathcal{O}_{\mathfrak{m}}\}$ and $v_{\mathfrak{m}}(S) = \min\{v_{\mathfrak{m}}(a) \mid a \in S\mathcal{O}_{\mathfrak{m}}\}$. For $0 \neq u \in \overline{R^{(n)}}$, u homogeneous, define $\langle u\rangle \in \mathfrak{P}(H)$ by $\langle u\rangle(\mathfrak{m}) = v_{\mathfrak{m}}(\{u\})$. If $U \in R^{(n)}$, U homogeneous and $\overline{U} = u \neq 0$, define $\langle U\rangle = \langle u\rangle$. For $0 \neq \frac{u}{v} \in \Omega(H)$ define $(\frac{u}{v}) \in \mathfrak{D}(H)$ so that $(\frac{u}{v}) = \langle u\rangle - \langle v\rangle$. For $D \in \mathfrak{P}(H)$ define $I(D) \subset R^{(n)}$ to be the homogeneous ideal generated by $\{U \in R^{(n)} \mid U$ homogeneous, $\overline{U} = 0$ or $\langle U\rangle \geq D\}$.

For $\mathfrak{m} \in \mathfrak{S}(H)$ we have the divisor $D_{\mathfrak{m}} \in \mathfrak{P}(H)$ defined by

$$D_{\mathfrak{m}}(\mathfrak{m}') = \begin{cases} 0 & \text{if } \mathfrak{m} \neq \mathfrak{m}' \\ 1 & \text{if } \mathfrak{m} = \mathfrak{m}'. \end{cases}$$

Clearly divisors of this type are in one-one correspondence with the elements of $\mathfrak{S}(H)$ and we also call $D_{\mathfrak{m}}$ a prime divisor. Note that a divisor is just a formal linear combination of prime divisors.

We shall write $R^{(2)} = k[X, Y, Z] = B$ and $R^{(3)} = k[X, Y, Z, W] = C$. Also, the underlying ring $k[X, Y, Z]$ of B without the graded structure shall be denoted by A. We have the identity map from A to (the underlying ring of) B, which shall be denoted by σ. Let $\mu : A \longrightarrow C$ defined by $\mu(h(X, Y, Z)) = W^r h(\frac{X}{W}, \frac{Y}{W}, \frac{Z}{W})$, where $r =$ degree of h.

We now fix, for the remainder of the notes, an irreducible $F \in A$ so that $f = \sigma(F) \in B_m$, $m \geq 3$ (i.e. F is homogeneous of degree m). It follows that $\mu(F) = F \in C$ and we shall assume that $\mu(F)$ defines an irreducible surface which is nonsingular in codimension 1. It is clear that then f defines an irreducible, nonsingular curve. Put $\Omega(\mu(F)) = \Omega$, $\Omega(f) = K$.

1.2 Pointsets on f

The effective divisors on f shall be called pointsets (with multiplicities) since f being one dimensional prime divisors on f are defined by points on f in the projective plane. We shall denote $\mathfrak{S}(f)$ by $\vartheta(f)$ = the set of points on f.

Let $L \subset B_n$ be a vector space. For a pointset $\mathcal{D}$ we define $\{L, \mathcal{D}\} = \{h \mid h \in L, \ \bar{h} = 0 \text{ or } \langle h\rangle \geq \mathcal{D}\}$ which is clearly a vector space again.

We also define $Fi(L, \mathcal{D})$ = the fixed part of L on $\mathcal{D}$ as follows.

If $h \in \{L, \mathcal{D}\}$ implies $\bar{h} = 0$, then $Fi(L, \mathcal{D}) = \mathcal{D}$. Otherwise $Fi(L, \mathcal{D})(M) = \min\{\langle h\rangle(M) \mid h \in \{L, \mathcal{D}\}, \ \bar{h} \neq 0\}$ for $M \in \vartheta(f)$.

A pointset $\mathcal{D}$ is called n-free if $Fi\{B_n, \mathcal{D}\} = \mathcal{D}$.

Remark 1: It is easy to see that if $\mathcal{D}$ is n-free and there exists $h \in \{B_n, \mathcal{D}\}$, $\bar{h} \neq 0$ then $\mathcal{D}$ is $(n + 1)$-free.

With each pointset $\mathfrak{D}$ we define degree of $\mathfrak{D} = \deg \mathfrak{D} = \sum \mathfrak{D}(M)$ where the sum is extended over all $M \in \vartheta(f)$. $\mathfrak{D}$ is called a simple pointset if $\mathfrak{D}(M) \le 1$ for all M; otherwise $\mathfrak{D}$ is singular.

A point $M \subset \overline{B}$, i.e. a minimal homogeneous prime ideal in $\overline{B}$, defines the minimal homogeneous prime ideal $M\overline{C} \subset \overline{C}$. Thus for each $\mathfrak{m} \in \mathfrak{P}(f)$ we have $\mathfrak{m}^* = M\overline{C} \in \mathfrak{P}(\mu(F))$. This induces a map from $\mathfrak{D}(f)$ into $\mathfrak{D}(\mu(F))$ which we denote by a^*. Any $D \in \mathfrak{D}(\mu(F))$ so that $D = \mathfrak{G}^*$ for $\mathfrak{G} \in \mathfrak{D}(f)$ shall be called a homogeneous divisor, since it consists of lines through $(0, 0, 0, 1)$ in the projective 3-space. It is clear that if $G \in A$ with $\sigma(G) = g \in B_n$ for some n then $\langle\mu G\rangle = \langle g\rangle^*$.

A pointset $\mathfrak{D}$ is called n-gg (n-geometrically generic) if $\dim_k \{B_n, \mathfrak{D}\} = \max\{0, \dim_k B_n - \deg \mathfrak{D}\}$. $\mathfrak{D}$ is called gg if $\mathfrak{D}$ is n-gg for $n = 0, 1, 2, \ldots, m - 1$. Note: $\dim_k B_n = \frac{1}{2}(n + 1)(n + 2)$.

1.3 Residuation on f

We shall prove the following residuation theorem.

Proposition 1: Let $\mathfrak{D}, \mathfrak{G}$ be pointsets so that $\mathfrak{D} = \langle g\rangle$, $0 \ne \overline{g} \in \overline{B}_t$, $\mathfrak{D} + \mathfrak{G} = \langle h\rangle$, $0 \ne h \in \overline{B}_s$. Then clearly $s \ge t$ and there exists $g' \in B_{s-t}$ so that $\langle g'\rangle = \mathfrak{G}$.

Proof: It is enough to prove that $\overline{h} = u\overline{g}$ for some $u \in \overline{B}$; for then it follows that u is homogeneous of degree $s - t$, $u \ne 0$ so that there exists $g' \in B_{s-t}$, $\overline{g}' = u$ and hence $\langle g'\rangle = \langle h\rangle - \langle g\rangle = \mathfrak{G}$.

Thus we want to show $\overline{h} \in \overline{g}\,\overline{B}$.

It is well-known that $\overline{g}$ being homogeneous all the associated primes of $\overline{g}$ are homogeneous, and by the unmixedness theorem all of them are minimal. [3] p. 203. (We apply the theorem to the ideal generated by g, f in B and then take residues modulo f.)

If $\mathfrak{n}$ is the primary component of $\overline{g}$ for an associated prime ideal $\mathfrak{m}$, then clearly $\overline{h} \in \mathfrak{n}$ if and only if $\overline{h}\,\overline{B}_{(\mathfrak{m})} \subset \mathfrak{n}\overline{B}_{(\mathfrak{m})}$. Thus it is enough to prove that for all the associated prime ideals $\mathfrak{m}$ of $\overline{g}$, $\overline{h}\,\overline{B}_{(\mathfrak{m})} \subset \mathfrak{n}\overline{B}_{(\mathfrak{m})}$. But $\overline{B}_{(\mathfrak{m})}$ being a valuation ring, $\overline{h}\,\overline{B}_{(\mathfrak{m})} \subset \mathfrak{n}\overline{B}_{(\mathfrak{m})}$ if and only if $\min\{v_{\mathfrak{m}}(u) \mid u \in \overline{h}\,\overline{B}_{(\mathfrak{m})}\} \ge \min\{v_{\mathfrak{m}}(u) \mid u \in \mathfrak{n}\overline{B}_{(\mathfrak{m})}\}$. Now $\overline{g}\,\overline{B} = \mathfrak{n} \cap \mathfrak{n}'$ where none of the associated prime ideals of $\mathfrak{n}'$ are contained in $\mathfrak{m}$ so that $\mathfrak{n}'\overline{B}_{(\mathfrak{m})} = B_{(\mathfrak{m})}$. Hence $\mathfrak{n}\overline{B}_{(\mathfrak{m})} = \overline{g}\,\overline{B}_{(\mathfrak{m})}$ and we simply get $\overline{h} \in \mathfrak{n}$ if and only if $v_{\mathfrak{m}}(\overline{h}) \ge v_{\mathfrak{m}}(\overline{g})$. This follows from $\langle h\rangle \ge \langle g\rangle$ and hence $\overline{h} \in \overline{g}\,B$.

Remark 2: Note that we have proved above $\overline{I(\langle g\rangle)} = \overline{g}\,\overline{B}$ whenever the only associated prime ideals of $\overline{g}\,\overline{B}$ are minimal and this result will hold in the general case when $\overline{B}$ is replaced by $\overline{R^{(n)}}$ as in 1.1.

1.4 The tangents at the origin

Let $G \in A$, $G \neq 0$. Then $\sigma(G) = \sum_{n=0}^{\infty} g_n$, $g_n \in B_n$ where $g_n = 0$ for all but finitely many values of n. Let $d = \min\{n \mid g_n \neq 0\}$. Then d is called the order of G and is denoted by $\mathrm{Ord}\, G$. We define $\ell(G) =$ the leading form of $G = g_d$. We define $\ell(0) = 0$.

For any set $S \subset A$ define $\ell(S) = \{\ell(H) \mid H \in S\}$. For any ideal $I \subset A$, $\ell(I)B$ is a homogeneous ideal in B. $\sigma^{-1}(\ell(I)B)$ defines the tangent cone at the origin of the affine variety defined by I. Suppose $F \in I$. Then $\ell(I)B$ also defines a pointset on f called the tangent set of I, $\mathfrak{F}$ say, so that $\mathfrak{F}(\mathfrak{m}) = \min\{v_{\mathfrak{m}}(\overline{u}) \mid u \in \ell(I)B$, u homogeneous$\}$. (Assume $\overline{\ell(I)B} \neq 0$.)

1.5 Divisors on $\mu(F)$ and the Bertini Theorem

Recall that two divisors of Δ, E on $\mu(F)$ are said to be linearly equivalent if $\Delta - E = (g)$ for some $g \in \Omega$. Let L be a collection of linearly equivalent effective divisors. L is called a linear system if, with fixed $\Delta_0 \in L$, the set of functions $\mathcal{L}' = \{h \in \Omega \mid h = 0$ or $(h) = \Delta - \Delta_0$ for some $\Delta \in L\}$ is a vector space over k. If $\mathcal{L} \subset \overline{C}_n$, $\mathcal{L} \neq 0$, is a vector space over k, we obtain a linear system putting

$$L = |\mathcal{L}| = \{\langle g\rangle \mid g \in \mathcal{L},\ g \neq 0\}.$$

(If we fix $\Delta_0 = \langle g_0\rangle$, $g_0 \in \mathcal{L}$, then $\mathcal{L}' = \{\frac{g}{g_0} \mid g \in \mathcal{L}\}$.)

In general, replacing Δ_0 by Δ_0' has the effect of replacing $\mathcal{L}'$ by $h_0\mathcal{L}'$ for some $0 \neq h_0 \in \Omega$. In particular, the subfield of Ω generated by $\{\frac{h_1}{h_2} \mid 0 \neq h_i \in \mathcal{L}'\}$ depends on L only and will be denoted by $k(L)$.

Given a linear system L, we define the divisor $\mathrm{Fi}(L)$, called the fixed component of L, by

$$\mathrm{Fi}(L)(\mathfrak{m}) = \min\{\Delta(\mathfrak{m}) \mid \Delta \in L\}$$

for any $\mathfrak{m} \in \mathfrak{S}(\mu(F))$. Then $L' = \{\Delta - \mathrm{Fi}(L) \mid \Delta \in L\}$ is again a linear system, $\mathrm{Fi}(L') = 0$ and $k(L) = k(L')$.

Bertini's theorem now states the following: If L' is a linear system without fixed component (i.e. $Fi(L') = 0$) and such that the transcendence degree of $k(L')$ over k is at least 2, then there exists $E \in L'$ such that E is a prime divisor. (This is true even for "most" members of L'.)

In fact, from the proof of Thm. 4.2 [2] p. 61 we can deduce that if every member of L' is reducible, i.e. not a prime divisor, then the transcendence degree of $k(L')$ over k is at most 1.

§2 THE POINTSET $\mathcal{R}$

In this section we shall prove

Propostition 2: There exists a simple pointset $\mathcal{R}$ on f so that $\deg \mathcal{R} = \frac{1}{2} m(m-1)$, $\mathcal{R}$ is $(m-2)$ gg and $(m-1)$ free, i.e. there is no $g \in B_{m-2}$ so that $\overline{g} \neq 0$, $\langle\overline{g}\rangle \geq \mathcal{R}$ and for any $M \in \mathcal{V}(f)$ there exists $g \in \{B_{m-1}, \mathcal{R}\}$, $\overline{g} \neq 0$ with $\langle g\rangle(M) = \mathcal{R}(M)$. In addition $\mathcal{R}$ is then m-free.

Proof: First we shall prove the following lemmas.

Lemma 1: Let $\mathcal{L} \subset B_n$, $n \leq m-1$ be a vector space and $\mathcal{B}$ a pointset. Also let $\mathcal{M}$ be any point. Then $\dim \{\mathcal{L}, \mathcal{B}\} - 1 \leq \dim \{\mathcal{L}, \mathcal{B} + \mathcal{M}\} \leq \dim \{\mathcal{L}, \mathcal{B}\}$.

Corollary 1: $\dim \{\mathcal{L}, \mathcal{B}\} \geq \dim \mathcal{L} - \deg \mathcal{B}$.

Corollary 2: If $Fi\,\{\mathcal{L}, \mathcal{B}\}(M) = \mathcal{B}(M)$ and $\dim \{\mathcal{L}, \mathcal{B}\} \neq 0$, then $\dim \{\mathcal{L}, \mathcal{B} + \mathcal{M}\} = \dim \{\mathcal{L}, \mathcal{B}\} - 1$.

Lemma 2: There exists a simple pointset $\mathcal{R}$ so that $\deg \mathcal{R} = \frac{1}{2} m(m-1)$ and $\mathcal{R}$ is gg.

Lemma 3: Let $\mathcal{J}$ be a pointset of degree $\frac{1}{2} m(m-1)$. Then there exists $g \in B_{m-1}$, $\overline{g} \neq 0$ with $\langle g\rangle = \mathcal{J} + \mathcal{J}'$ where $\mathcal{J}'$ is a pointset of degree $\frac{1}{2} m(m-1)$. Also $\mathcal{J}$ is $(m-2)$ gg if and only if $\mathcal{J}'$ is $(m-2)$ gg.

Lemma 4: Let $\mathcal{R}$ be a simple pointset of degree $\frac{1}{2} m(m-1)$ which is $(m-2)$ gg. Then $\mathcal{R}$ is $(m-1)$ free.

Corollary 3: $\mathcal{R}$ is m-free.

The proof of Proposition 2 will clearly follow from Lemmas 3, 4 and Corollary 3.

Note: We will use without further comment that if $0 \neq h \in B_d$ and $d < m$ then $\overline{h} \neq 0$ since degree of $f = m$.

Proof Lemma 1: If $\dim\{\mathcal{L}, \mathcal{D}\} = 0$ then $\dim\{\mathcal{L}, \mathcal{D} + \mathcal{M}\} = 0$ and there is nothing to prove. If $\dim\{\mathcal{L}, \mathcal{D}\} > 0$ then clearly there exists $h \in \{\mathcal{L}, \mathcal{D}\}$, $h \neq 0$. Since $\mathcal{L} \subset B_n$ with $n < m$, $\overline{h} \neq 0$ and $\langle h\rangle \geq \mathcal{D}$.

Clearly $\{\mathcal{L}, \mathcal{D} + \mathcal{M}\} \subset \{\mathcal{L}, \mathcal{D}\}$ and $\dim\{\mathcal{L}, \mathcal{D} + \mathcal{M}\} \leq \dim\{\mathcal{L}, \mathcal{D}\}$. To prove the second inequality it is enough to consider the case when $\dim\{\mathcal{L}, \mathcal{D} + \mathcal{M}\} < \dim\{\mathcal{L}, \mathcal{D}\}$ so that there is $h \in \{\mathcal{L}, \mathcal{D}\}$, $\overline{h} \neq 0$, $\langle h\rangle \geq \mathcal{D}$, $\langle h\rangle \ngeq \mathcal{D} + \mathcal{M}$. Then $\mathcal{D}(\mathcal{M}) \leq \langle h\rangle(\mathcal{M}) < (\mathcal{D} + \mathcal{M})(\mathcal{M}) = \mathcal{D}(\mathcal{M}) + 1$, i.e. $\langle h\rangle(\mathcal{M}) = \mathcal{D}(\mathcal{M})$.

Let $g \in \{\mathcal{L}, \mathcal{D}\}$, $g \neq 0$. Then $\langle g\rangle \geq \mathcal{D}$. If $\langle g\rangle(\mathcal{M}) \geq \mathcal{D}(\mathcal{M}) + 1$ then $g \in \{\mathcal{L}, \mathcal{D} + \mathcal{M}\}$. Otherwise since $\langle g\rangle(\mathcal{M}) \geq \mathcal{D}(\mathcal{M})$, we have $\langle g\rangle(\mathcal{M}) = \mathcal{D}(\mathcal{M}) = \langle h\rangle(\mathcal{M})$. Hence there exists $\lambda \in k$ so that $\langle g + \lambda h\rangle(\mathcal{M}) > \mathcal{D}(\mathcal{M}) = \langle g\rangle(\mathcal{M}) = \langle h\rangle(\mathcal{M})$ so that $g + \lambda h \in \{\mathcal{L}, \mathcal{D} + \mathcal{M}\}$. Thus $\{\mathcal{L}, \mathcal{D}\}$ is generated as a vector space by h and $\{\mathcal{L}, \mathcal{D} + \mathcal{M}\}$. This proves the result.

Proof Corollary 1: Follows by induction on $\deg \mathcal{D}$ using the inequality

$$\dim\{\mathcal{L}, \mathcal{D}\} - 1 \leq \dim\{\mathcal{L}, \mathcal{D} + \mathcal{M}\}.$$

Proof Corollary 2: Since $\mathrm{Fi}\,\{\mathcal{L}, \mathcal{D}\}(\mathcal{M}) = \mathcal{D}(\mathcal{M})$ and $\dim\{\mathcal{L}, \mathcal{D}\} \neq 0$ there exists $h \in \{\mathcal{L}, \mathcal{D}\}$, $\overline{h} \neq 0$ so that $\langle h\rangle(\mathcal{M}) = \mathcal{D}(\mathcal{M})$. Then $\{\mathcal{L}, \mathcal{D} + \mathcal{M}\} \subsetneqq \{\mathcal{L}, \mathcal{D}\}$. Thus from Lemma 1 it follows that $\dim\{\mathcal{L}, \mathcal{D}\} - 1 \leq \dim\{\mathcal{L}, \mathcal{D} + \mathcal{M}\} < \dim\{\mathcal{L}, \mathcal{D}\}$, hence the result.

Proof 1 Lemma 2: We shall prove the following more general result.

There exists a simple pointset $\mathcal{D}_t$ of degree t, $t = 0, 1, \ldots$ so that $\mathcal{D}_t$ is gg.

Put $\mathcal{D}_0 = 0$ which clearly satisfies the conditions. Assume

$\mathfrak{D}_0, \mathfrak{D}_1, \ldots, \mathfrak{D}_{t-1}$ are already defined. Put $V_{s,d} = \{B_d, \mathfrak{D}_s\}$ and $n_{s,d} = \dim V_{s,d}$ $0 \le s \le t-1$. We shall find a point $\mathfrak{m}$ so that $\mathfrak{D}_t = \mathfrak{D}_{t-1} + \mathfrak{m}$ is simple and gg. For each $V_{(t-1),d}$, $1 \le d < m$ such that $V_{(t-1),d} \ne \{0\}$ choose $g_d \in V_{(t-1),d}$ so that $g_d \ne 0$ so that $\overline{g}_d \ne 0$.

Let N_t be the set of all such g_d. Choose $\mathfrak{m}$ such that $g \in N_t$ implies $\langle g\rangle(\mathfrak{m}) = 0$ and such that $\mathfrak{D}_{(t-1)}(\mathfrak{m}) = 0$. Put $\mathfrak{D}_t = \mathfrak{D}_{t-1} + \mathfrak{m}$. We claim that if $V_{t,d} = \{B_d, \mathfrak{D}_t\}$, $n_{t,d} = \dim V_{t,d}$ then

$$n_{t,d} = \max\{0, \dim B_d - \deg \mathfrak{D}_t\}.$$

If $n_{(t-1),d} = 0$, then clearly $n_{t,d} = 0$. Also $n_{(t-1),d} = 0$ implies by induction hypothesis

$$0 = \max\{0, \dim B_d - \deg \mathfrak{D}_{(t-1)}\}$$

so that $\dim B_d - \deg \mathfrak{D}_{(t-1)} \le 0$. Hence

$$\dim B_d - \deg \mathfrak{D}_t \le 0$$

and

$$n_{t,d} = \max\{0, \dim B_d - \deg \mathfrak{D}_t\}.$$

If $n_{(t-1),d} \ge 1$, then by induction hypothesis

$$n_{(t-1),d} = \dim B_d - \deg \mathfrak{D}_{(t-1)} \ge 1.$$

Also by Corollary 2, $n_{t,d} = n_{(t-1),d} - 1$ since by the choice of $\mathfrak{m}$ there exists $g_d \in V_{(t-1),d}$, $g_d \ne 0$ and $\langle g_d\rangle(\mathfrak{m}) = 0$ so that $\mathrm{Fi}(V_{(t-1),d})(\mathfrak{m}) = 0 = \mathfrak{D}_{(t-1)}(\mathfrak{m})$. Hence

$$n_{t,d} = \dim B_d - \deg \mathfrak{D}_t \ge 0.$$

Proof 2 Lemma 2: It is enough to prove that some affine piece of the curve f contains a pointset having the desired properties. We choose the piece outside the line $Z = 0$. (Note $Z \nmid f$.) This piece, denoted $\vartheta'(f)$, has affine coordinate ring $k[X', Y']/f'(X', Y') = k[\overline{X}', \overline{Y}']$ where $X' = \frac{X}{Z}$, $Y' = \frac{Y}{Z}$ and $f'(X', Y') = f(X', Y', 1)$. An element $h \in B_d$ induces a k-valued function h' on $\vartheta'(f)$ defined for $\mathfrak{m} \in \vartheta'(f)$ by $h'(\mathfrak{m}) = \mathfrak{m}$ - residue of $h(\overline{X}', \overline{Y}', 1)$.

Now $\deg f' = \deg f = m$, and hence for $d < m$, $B_d \simeq B'_d = \{h' \mid h \in B_d\}$, a vector space of k-valued functions on $\vartheta'(f)$ of dimension $s(d) = \frac{1}{2}(d+1)(d+2)$. Let $w_1, \ldots, w_{s(d)}$ be monomials of degree $\leq d$ in X' and Y'. They form a basis for B'_d. Let $h \in B_d$ and $\mathfrak{m} \in \vartheta(f')$. Then clearly $\langle h\rangle \geq \mathfrak{m}$ if and only if $h'(\mathfrak{m}) = 0$. Thus for simple $\mathcal{R} = \mathfrak{m}_1 + \cdots + \mathfrak{m}_t$, $\mathfrak{m}_i \in \vartheta'(f)$, and $h \in B_d$, $\langle h\rangle \geq \mathcal{R}$ if and only if

$$h'(\mathfrak{m}_i) = 0, \quad i = 1, 2, \ldots, t.$$

It follows that $\{B_d, \mathcal{R}\}$ is isomorphic to the space of solutions of

$$\sum_{i=1}^{s(d)} \lambda_i w_i(\mathfrak{m}_j) = 0, \quad j = 1, \ldots, t.$$

To find $\mathcal{R}$ with the required properties it is enough to determine $\mathfrak{m}_1, \ldots, \mathfrak{m}_t$ ($t = \frac{1}{2}m(m-1)$, all $\mathfrak{m}_i$ distinct) so that the matrix $(w_i(\mathfrak{m}_j))$ has maximal possible rank $\min\{t, s(d)\}$ for $d = 0, 1, \ldots, m-1$. For then

$$\begin{aligned}\dim\{B_d, \mathcal{R}\} &= s(d) - \operatorname{rank}(w_i(\mathfrak{m}_j)) \\ &= s(d) - \min\{t, s(d)\} \\ &= \max\{0, s(d) - t\}\end{aligned}$$

and $\mathcal{R}$ is gg.

We now appeal to the following elementary lemma: Let S be a set and V a vector space of k-valued functions on S, k a field. If $\dim V \geq t$, then there exist $\mathfrak{m}_1, \ldots, \mathfrak{m}_t \in S$ and $\mu_1, \ldots, \mu_t \in V$ such that $\mu_i(\mathfrak{m}_j) = \delta_{ij}$.

Applying this to $S = \vartheta'(f)$ and $V = B_{m-2}$, we find $\mathfrak{m}_1, \ldots, \mathfrak{m}_t$ so that the matrix $(w_i(\mathfrak{m}_j))$, $i, j = 1, \ldots, t = s(m-2)$ is invertible. Then clearly for any $d \leq m-1$, the matrix $(w_i(\mathfrak{m}_j))$, $i = 1, \ldots, s(d)$, $j = 1, \ldots, t$ has maximal possible rank, as required.

<u>Proof Lemma 3</u>: By Corollary 1 $\dim\{B_{m-1}, \mathfrak{d}\} \geq m$ so that there exists $0 \neq g \in \{B_{m-1}, \mathfrak{d}\}$ with $\langle g\rangle \geq \mathfrak{d}$ and since by Bezout's theorem $\deg\langle g\rangle = m(m-1)$ we have $\langle g\rangle = \mathfrak{d} + \mathfrak{d}'$ with $\deg \mathfrak{d}' = \frac{1}{2}m(m-1)$.

It is enough to prove that if $\mathfrak{d}$ is not $(m-2)$ gg then $\mathfrak{d}'$

is not $(m-2)$ gg, the result being symmetric in $\mathcal{S}, \mathcal{S}'$.

Thus let $h \in B_{m-2}$, $h \neq 0$ and $\langle h\rangle = \mathcal{S} + \mathcal{T}$ for some pointset $\mathcal{T}$. Then by Bezout's theorem $\deg \mathcal{T} = m(m-2) - \frac{1}{2}m(m-1) = \frac{m(m-3)}{2}$ $= \dim(B_{m-3}) - 1$ so that $\dim\{B_{m-3}, \mathcal{T}\} \geq 1$ by Corollary 1. Hence there exists $0 \neq h' \in \{B_{m-3}, \mathcal{T}\}$ so that $\overline{h}' \neq 0$ and $\langle h'\rangle = \mathcal{T} + \mathcal{T}'$.

Then $\langle gh'\rangle = \mathcal{S} + \mathcal{T} + \mathcal{S}' + \mathcal{T}' = \langle h\rangle + \mathcal{S}' + \mathcal{T}'$ and by Proposition 1 there exists $g' \in B_{m-2}$ so that $\overline{g}' \neq 0$ and $\langle g'\rangle = \mathcal{S}' + \mathcal{T}'$. Hence $\dim\{B_{m-2}, \mathcal{S}'\} \neq 0$ and $\mathcal{S}'$ is not $(m-2)$ gg.

<u>Proof Lemma 4</u>: We have to show that for any $\mathfrak{m} \in \vartheta(f)$ there exists $g \in \{B_{m-1}, \mathcal{R}\}$ so that $\langle g\rangle(\mathfrak{m}) = R(\mathfrak{m})$.

First consider the case when $R(\mathfrak{m}) \neq 0$ and hence $R(\mathfrak{m}) = 1$. Then we may write $\mathcal{R} = (\mathcal{R} - \mathfrak{m}) + \mathfrak{m}$ where $(\mathcal{R} - \mathfrak{m}) \in \mathfrak{B}(f)$. Then $\dim\{B_{m-2}, \mathcal{R} - \mathfrak{m}\} \geq 1$ by Corollary 1 and $\dim\{B_{m-2}, \mathcal{R}\} = 0$ by hypothesis. Hence there exists $h_{\mathfrak{m}} \in \{B_{m-2}, \mathcal{R} - \mathfrak{m}\}$ so that $h_{\mathfrak{m}} \notin \{B_{m-2}, \mathcal{R}\}$ and hence $\langle h_{\mathfrak{m}}\rangle = (\mathcal{R} - \mathfrak{m}) + \mathcal{T}$ for some $\mathcal{T}$ with $\langle h_{\mathfrak{m}}\rangle(\mathfrak{m}) = 0$. Choosing $h'_{\mathfrak{m}} \in B_1$ so that $\langle h'_{\mathfrak{m}}\rangle(\mathfrak{m}) = 1$, we have $h_{\mathfrak{m}}h'_{\mathfrak{m}} \in \{B_{m-1}, \mathcal{R}\}$ with $\langle h_{\mathfrak{m}}h'_{\mathfrak{m}}\rangle(\mathfrak{m}) = 1 = R(\mathfrak{m})$.

Now let $\mathfrak{m}$ be a point with $R(\mathfrak{m}) = 0$. Fix $0 \neq g \in \{B_{m-1}, \mathcal{R}\}$ and write $\langle g\rangle = \mathcal{R} + \mathcal{R}'$. (Existence of g follows as in Lemma 3.) If $R'(\mathfrak{m}) = 0$ then $\langle g\rangle(\mathfrak{m}) = 0 = R(\mathfrak{m})$ and we have nothing to prove. Thus we may assume that $R'(\mathfrak{m}) \geq 1$. For each $\mathfrak{n}$ with $R(\mathfrak{n}) = 1$ choose $h_{\mathfrak{n}} \in \{B_{m-2}, \mathcal{R} - \mathfrak{n}\}$ as above so that $\langle h_{\mathfrak{n}}\rangle(\mathfrak{n}) = 0$. If for some $\mathfrak{n}$, $\langle h_{\mathfrak{n}}\rangle(\mathfrak{m}) = 0$, we may choose $h_{\mathfrak{m}\mathfrak{n}} \in B_1$ with $\langle h_{\mathfrak{m}\mathfrak{n}}\rangle(\mathfrak{m}) = 0$ and $\langle h_{\mathfrak{m}\mathfrak{n}}\rangle(\mathfrak{n}) = 1$. Then $h_{\mathfrak{n}}h_{\mathfrak{m}\mathfrak{n}} \in \{B_{m-1}, \mathcal{R}\}$ and $\langle h_{\mathfrak{n}}h_{\mathfrak{m}\mathfrak{n}}\rangle(\mathfrak{m}) = 0 = R(\mathfrak{m})$ as required.

Thus we have only to consider the case when $\langle h_{\mathfrak{n}}\rangle(\mathfrak{m}) \geq 1$ for all $\mathfrak{n}$ with $\mathcal{R}(\mathfrak{n}) = 1$. We shall deduce a contradiction from this, thus proving the lemma.

We have

$$\langle h_{\mathfrak{n}}\rangle = (\mathcal{R} - \mathfrak{n}) + \mathfrak{m} + \mathcal{T}_{\mathfrak{n}}.$$

Hence $\deg \mathcal{T}_{\mathfrak{n}} = \dim(B_{m-3}) - 1$ and as in Lemma 3 there exists $h'_{\mathfrak{n}} \in \{B_{m-3}, \mathcal{T}_{\mathfrak{n}}\}$ so that

$$\langle h'_{\mathfrak{n}}\rangle = \mathcal{T}_{\mathfrak{n}} + \mathcal{T}'_{\mathfrak{n}}.$$

Also $\langle g\rangle = (\mathcal{R} - \mathfrak{n}) + \mathfrak{m} + (\mathcal{R}' - \mathfrak{m}) + \mathfrak{n}$ so that

$$\langle gh'_{\mathfrak{n}}\rangle = \langle h_{\mathfrak{n}}\rangle + \mathcal{R}' - \mathfrak{m} + \mathfrak{n} + \mathcal{T}'_{\mathfrak{n}}.$$

By Proposition 1 there exists $g'_{\eta} \in \{B_{m-2}, \mathcal{R}' - \mathfrak{m}\}$ so that

$$\langle g'_{\eta}\rangle = \mathcal{R}' - \mathfrak{m} + \eta + \mathcal{T}'_{\eta}.$$

By Lemma 2, $\dim\{B_{m-2}, \mathcal{R}'\} = 0$. Since $\dim\{B_{m-2}, \mathcal{R}' - \mathfrak{m}\} \neq 0$, $\dim\{B_{m-2}, \mathcal{R}' - \mathfrak{m}\} = 1$ by Lemma 1, and g'_{η} is uniquely determined by η (up to a constant in k). So we may take $g'_{\eta} = g'$ for all η.

But then $\langle g'\rangle(\eta) \geq 1$ for all η with $\mathcal{R}(\eta) = 1$, that is $\langle g'\rangle \geq \mathcal{R}$ and we get $\dim\{B_{m-2}, \mathcal{R}\} \geq 1$, a contradiction.

Proof Corollary 3: As in Lemma 3 there exists $0 \neq g \in \{B_{m-1}, \mathcal{R}\}$, so that by Remark 1 $\mathcal{R}$ is m free.

§3 PROOF OF THE THEOREM

Proposition 3: Let $\mathcal{R}$ be as in Proposition 2. Let $0 \neq G \in A$ so that $\sigma(G) = g \in \{B_{m-1}, \mathcal{R}\}$ and write $\langle g\rangle = \mathcal{R} + \mathcal{R}'$. Then there exists $H \in A$ so that $\sigma(H) \in B_m$ and $(F, G + H)A = \mathfrak{A} \cap P$ where P is a prime ideal and $\sigma(\mathfrak{A}) = I(\mathcal{R})$. Further we can choose G so that $\mathcal{R}$ and $\mathcal{R}'$ have no common points, i.e. $\mathcal{R}(\mathfrak{m}) > 0$ implies $\mathcal{R}'(\mathfrak{m}) = 0$.

Proof: Consider the vector space

$$\mathcal{L} = \{\lambda\overline{G}\,\overline{W} + \overline{H} \mid H \in A \text{ with } \sigma(H) \in \{B_m, \mathcal{R}\}\} \subset \overline{C}_m.$$

We claim that $\mathrm{Fi}(|\mathcal{L}|) = \mathcal{R}^*$. Obviously $\mathrm{Fi}(|\mathcal{L}|) \geq \mathcal{R}^*$. On the other hand there exists $0 \neq \overline{\mu(H)} \in \mathcal{L}$ and $\langle\overline{\mu(H)}\rangle = \langle h\rangle^*$ ($h = \sigma(H) \in B_m$). Put differently, if $\mathbf{m}$ is not homogeneous $\langle\overline{\mu(H)}\rangle(\mathbf{m}) = 0$. Hence $\mathrm{Fi}(|\mathcal{L}|)$ is homogeneous and since $\mathcal{R}$ is m-free by Corollary 3, clearly $\mathcal{R}^* \geq \mathrm{Fi}(|\mathcal{L}|)$.

Now $\overline{G}\,\overline{W}$, $\overline{G}\,\overline{X}$, $\overline{G}\,\overline{Y}$, $\overline{G}\,\overline{Z} \in \mathcal{L}$ so that

$$\frac{\overline{X}}{\overline{W}}, \frac{\overline{Y}}{\overline{W}}, \frac{\overline{Z}}{\overline{W}} \in k(|\mathcal{L}|)$$

and $k(|\mathcal{L}|) = \Omega$ so that transcendence degree of $k(|\mathcal{L}|)$ over k is 2.

By Bertini's Theorem there exists a prime divisor $\mathbf{m} \in |\mathcal{L}|'$, i.e. for some $\lambda\overline{G}\,\overline{W} + \overline{H} \in \mathcal{L}$

$$\langle\lambda\overline{G}\,\overline{W} + \overline{H}\rangle = \mathcal{R}^* + \mathbf{m}.$$

With each divisor $\mathfrak{D} = \sum_{i=1}^{s} n_i \mathfrak{m}_i$ on $\mu(F)$ we can associate $\deg \mathfrak{D} = \sum n_i \deg \mathfrak{m}_i$ where $\deg \mathfrak{m}_i$ = degree of the irreducible curve on $\mu(F)$ defined by the ideal $\mathfrak{m}_i$. Then Bezout's theorem for surfaces yields for $U \in \overline{C}_n$, $U \neq 0$

$$\deg \langle U \rangle = nm.$$

Thus $\deg \langle \lambda \overline{G}\,\overline{W} + \overline{H} \rangle = m^2$. Also for any $\mathcal{M} \in \vartheta(f)$, $\deg (\mathcal{M}^*) = 1$, i.e. the degree of a homogeneous prime divisor is 1. Hence clearly $\deg \mathcal{R}^* = \deg \mathcal{R} = \frac{1}{2} m(m-1)$ so that $\deg (\mathfrak{m}) = \frac{1}{2} m(m+1) > 1$. In particular $\mathfrak{m}$ is not a homogeneous prime divisor. Hence we can conclude that $\lambda \neq 0$ since otherwise $\mathfrak{m}$ is homogeneous. Also $H \neq 0$ since otherwise $\mathfrak{m} = \mathcal{R}'^* + \langle W \rangle$ would not be prime.

Now $\lambda \overline{G}\,\overline{W} + \overline{H} \not\in \overline{W}\,\overline{C}$ since $H \in A$ and hence $\overline{W}$ does not belong to the prime ideal of $\mathfrak{m}$.

As in Remark 2 we can conclude that $(\lambda \overline{G}\,\overline{W} + \overline{H})\overline{C} = \overline{I(\mathcal{R}^*)} \cap \overline{I(\mathfrak{m})}$ and since $\overline{W}$ does not divide $\lambda \overline{G}\,\overline{W} + \overline{H}$ we have in affine 3-space

$$(\lambda G + H, F)A = \mathfrak{A} \cap P$$

where $\mu(P) = I(\mathfrak{m})$, $\mu(\mathfrak{A}) = I(\mathcal{R}^*)$. It is straightforward to check that $\sigma(\mathfrak{A}) = I(\mathcal{R})$.

Since $\lambda \neq 0$ we may take $\lambda = 1$ and we have the required $G + H$.

For the last remark we simply have to observe that for each $\mathcal{M}$ with $R(\mathcal{M}) = 1$ there is $0 \neq g_{\mathcal{M}} \in \{B_{m-1}, \mathcal{R}\}$ so that $\langle g_{\mathcal{M}} \rangle(\mathcal{M}) = 1$. We may then take a linear combination $\sum_{R(\mathcal{M})=1} C_{\mathcal{M}} g_{\mathcal{M}} = g$ so that $C_{\mathcal{M}} \in k$ and $\langle g \rangle(\mathcal{M}) = \langle g_{\mathcal{M}} \rangle(\mathcal{M}) = 1$. It is clear that $G = \sigma^{-1}(g)$ is the required element.

<u>Proposition 4</u>: For P as in Proposition 3 the tangent set $\mathfrak{D}$ of P satisfies $\mathfrak{D} \geq \mathcal{R}'$.

<u>Proof</u>: Let $F' \in P$. It is enough to prove that if $\overline{\ell(F')} \neq 0$ then $\langle \ell(F') \rangle \geq \mathcal{R}'$. Choose $F'' \in A$ so that $\sigma(F'') \in \{B_{m-1}, \mathcal{R}\}$ with $\overline{\sigma(F'')} \neq 0$ and $\langle \sigma(F'') \rangle(\mathcal{M}) = 0$ for all $\mathcal{M}$ with $\mathcal{R}'(\mathcal{M}) > 0$, i.e. $\sigma(F'')$ does not pass through any point of $\mathcal{R}'$. This is possible since, as $\mathrm{Fi}\{B_{m-1}, \mathcal{R}\} = \mathcal{R}$, for each $\mathcal{M}$ with $\mathcal{R}'(\mathcal{M}) > 0$ we find $F''_{\mathcal{M}}$ so that $\langle \sigma(F''_{\mathcal{M}}) \rangle(\mathcal{M}) = 0$ and $\sigma(F''_{\mathcal{M}}) \in \{B_{m-1}, \mathcal{R}\}$. A suitable

(general) linear combination of the $F''_{\mathfrak{m}}$'s will serve as F''.

We claim that $F'' \notin P$ since otherwise $\langle F''\rangle(m) > 0$; but $\langle F''\rangle = \langle \mu(F'')\rangle$ is homogeneous and m is not.

Now $F'F'' \in P \cap \mathfrak{A} = (F, G + H)A$. We claim that the tangent set of $(F, G + H)A$ is $\mathcal{R} + \mathcal{R}'$. Assuming this we find that $\langle \ell(F')\sigma(F'')\rangle \geq \mathcal{R} + \mathcal{R}'$. But $\langle \sigma(F'')\rangle(\mathfrak{m}) = 0$ if $\mathcal{R}'(\mathfrak{m}) > 0$ and hence $\langle \ell(F')\rangle(\mathfrak{m}) \geq \mathcal{R}'(\mathfrak{m})$. This means $\langle \ell(F')\rangle \geq \mathcal{R}'$, as required.

Since $\mathcal{R} + \mathcal{R}' = \langle g\rangle$ our claim is equivalent to $\ell((F, G + H)A)B = (f, g)B$ where $g = \sigma(G)$. Clearly $f, g \in \ell(F, G + H)A)B$. Also if $U_1F + U_2(G + H) \in (F, G + H)A$, we may as well assume that $\overline{\ell(U_2)} \neq 0$. Then $\ell(U_1)f + \ell(U_2)g \neq 0$ since otherwise $\overline{\ell(U_2)}\,\overline{g} = 0$ and neither $\overline{\ell(U_2)}$ nor $\overline{g}$ is zero. It is then clear that

$$\ell(U_1F + U_2(G + H)) = \ell(U_1)f + \ell(U_2)g$$

so that

$$\ell((F, G + H)A)B = (f, g)B.$$

Corollary 4: $P \subset ((X, Y, Z)A)^{m-1}$.

Proof: Let $F' \in P$, $F' \neq 0$. Then $\ell(F') \neq 0$ and either $\overline{\ell(F')} = 0$ and hence $\mathrm{Ord}(F') \geq m$ or $\langle \ell(F')\rangle \geq \mathcal{R}'$. In the latter case, $\mathcal{R}'$ being $(m - 2)$ gg, $\ell(F') \in B_d$ with $d \geq m - 1$ so that $\mathrm{Ord}(F') \geq (m - 1)$.

Proposition 5: For $\mathcal{R}, \mathcal{R}'$, P as before, $\{B_{m-1}, \mathcal{R}'\} \subset \ell(P)$ and in fact if $U_1, \ldots, U_s$ generate P, then those of the $\ell(U_i)$ which are in B_{m-1} generate $\{B_{m-1}, \mathcal{R}'\}$.

Proof: For the first statement it is enough to prove that for $g_1 \in \{B_{m-1}, \mathcal{R}'\}$, $g_1 \neq 0$ there exists $h_1 \in B_m$ so that $\sigma^{-1}(g_1 + h_1) \in P$, for then $g_1 = \ell(\sigma^{-1}(g_1 + h_1)) \in \ell(P)$.

Now $0 \neq g_1 \in B_{m-1}$ and $\langle g_1\rangle = \mathcal{R}' + \mathcal{J}$. Also $\langle g\rangle = \mathcal{R} + \mathcal{R}'$. Also if $h = \sigma(H)$, we have by construction $\langle h\rangle = \mathcal{R} + \mathcal{J}'$ for some pointset $\mathcal{J}'$. Now $\langle hg_1\rangle = \langle g\rangle + \mathcal{J} + \mathcal{J}'$ and by Proposition 1 there exists $0 \neq h_1 \in B_m$ with $\langle h_1\rangle = \mathcal{J} + \mathcal{J}'$. Then

$$\overline{hg_1} = \overline{gh_1}$$

so that

$$HG_1 = GH_1 + UF$$

where

$$G_1 = \sigma^{-1}(g_1), \quad H_1 = \sigma^{-1}(G).$$

Hence

$$(G + H)G_1 = G(G_1 + H_1) + UF.$$

Now $G + H$, $F \in P$ so that $G(G_1 + H_1) \in P$. Also $G \notin P$ (for the same reason as $F'' \notin P$ in the proof of Proposition 4). Hence $G_1 + H_1 \in P$ as required.

For the second statement observe that given a set of generators $U_1, \ldots, U_s$ of P, any $0 \neq g_1 \in \{B_{m-1}, \mathcal{R}'\}$ is of the form $\ell(a_1U_1 + \cdots + a_sU_s)$, $a_1, \ldots, a_s \in A$.

Now $\ell(U_i) \in B_r$, $r \geq (m - 1)$ by Corollary 4. Since $0 \neq g_1 \in B_{m-1}$,

$$g_1 = \sum_{\substack{\ell(U_i) \in B_{m-1} \\ \ell(a_i) \in k}} \ell(a_i)\ell(U_i).$$

This proves the second statement.

Proof of the Theorem: By Corollary 1 $\dim\{B_{m-1}, \mathcal{R}'\} \geq m$. It follows that the prime ideal P needs at least m generators.

§4 SOME QUESTIONS

Remark 3: If the pointset $\mathcal{R}'$ in §3 could be proved to be $(m - 1)$ free then the proof of Proposition 4 could be sharpened to prove that the tangent set of P is $\mathcal{R}'$.

Remark 4: One way to prove that $\mathcal{R}'$ is $(m - 1)$ free is to show that $\mathcal{R}'$ can be arranged to be a simple pointset disjoint from $\mathcal{R}$ (i.e. so that $\mathcal{R}'(\mathfrak{m}) \geq 1$ implies $\mathcal{R}(\mathfrak{m}) = 0$). Then one may apply Lemma 4 to $\mathcal{R}'$ instead of $\mathcal{R}$. (By Lemma 3, $\mathcal{R}'$ is $(m - 2)$ gg since $\mathcal{R}$ is $(m - 2)$ gg.) This could be done in characteristic 0 by applying Bertini's theorem on variable singularities to the linear system of pointsets $\{\langle g\rangle - \mathcal{R} \mid g \in \{B_{m-1}, \mathcal{R}\},\ g \neq 0\}$. We ask the following question in general:

Q 1: Does there exist $g \in B_{m-1}$ so that $\langle g\rangle = \mathcal{R} + \mathcal{R}'$, $\deg \mathcal{R} = \deg \mathcal{R}' = \frac{1}{2}m(m - 1)$ and both $\mathcal{R}$, $\mathcal{R}'$ are $(m - 2)$ gg, simple and disjoint?

Q 2: Let $g \in B_{m-1}$ be a generic member and write $\langle g \rangle = \mathcal{R} + \mathcal{R}'$, $\deg \mathcal{R} = \deg \mathcal{R}' = \frac{1}{2} m(m-1)$. Does it follow that

i) $\mathcal{R}$ is $(m-2)$ gg

ii) $\mathcal{R}$ is $(m-1)$ free?

Another possible way would be to have a lemma of the type of Lemma 2. We ask

Q 3: Let $g \in B_{m-1}$ with $\langle g \rangle = \mathcal{R} + \mathcal{R}'$. Given that $\mathcal{R}$ is $(m-1)$ free, does it follow that $\mathcal{R}'$ is $(m-1)$ free?

Clearly we may assume, if necessary, that $\mathcal{R}$ is $(m-2)$ gg or gg in the above question.

Remark 5: As in Remark 3, if $\mathcal{R}'$ is $(m-1)$ free, then $\mathcal{R}'$ is the tangent set of P. Thus if we could arrange $\mathcal{R}'$ to be of the type $\mathcal{R}' = \frac{1}{2} m(m-1)\mathfrak{m}$ the resulting curve then would have only one tangent (set theoretically) at the origin. Hence we ask

Q 4: Can we arrange $\mathcal{R}'$ to be $(m-1)$ free and of the form $\mathcal{R}' = \frac{1}{2} m(m-1)\mathfrak{m}$?

Remark 6: Just as we defined pointsets on a curve f we may define pointsets in the plane (without multiplicities) as formal sums of distinct points. Every simple pointset on f clearly corresponds to such a pointset in the plane.

Thus to construct $\mathcal{R}$ we may start with a set $\tilde{\mathcal{R}}$ of $\frac{1}{2} m(m-1)$ points in the plane so that there is no curve of degree $m-2$ passing through $\tilde{\mathcal{R}}$ and the curves of degree $(m-1)$ through $\tilde{\mathcal{R}}$ have no fixed point outside $\tilde{\mathcal{R}}$. If we could pass a nonsingular curve f through $\tilde{\mathcal{R}}$, then the image of $\tilde{\mathcal{R}}$ on f would be the required set $\mathcal{R}$ on f. Thus in general we ask:

Q 5: For a set $\tilde{\mathcal{R}}$ as above, what is $\min\{n \mid$ there is a nonsingular curve of degree n through $\tilde{\mathcal{R}}\}$?

In characteristic 0, using Lemma 4 and a type of descending induction one can prove that indeed the minimum is $(m-1)$. In characteristic not equal to zero the question remains open.

One has also a more generalized version of Q 5, viz.

Q 6: For a pointset $\mathcal{S}$ in the plane how to find $\min\{n \mid$ there is a nonsingular curve of degree n through $\mathcal{S}\}$?

BIBLIOGRAPHY

[1] Macaulay, "The Algebraic Theory of Modular Systems", Stechert Hafner Service Agency (Reprint), 1964.

[2] Matsusaka, Memoirs of the College of Science, Univ. of Kyoto, Ser A, Vol. XXVI, 1950, p. 51-52.

[3] Zariski, Samuel, "Commutative Algebra" Vol. II, Van Nostrand Company.

PRIME IDEALS IN POWER SERIES RINGS

Jimmy T. Arnold

Virginia Polytechnic Institute and State University

ABSTRACT. In this paper we wish to briefly review some known results concerning the ideal structure of the formal power series ring R[[X]]. As the title indicates, primary consideration will be given to prime ideals in R[[X]]. We begin by discussing some of the basic difficulties which arise in relating the ideal structure of R[[X]] with that of R. We then consider the Krull dimension of R[[X]] and, finally, we review some results on valuation overrings of D[[X]], where D is an integral domain.

1. Introduction. Our notation and terminology are essentially that of [8]. Throughout, R denotes a commutative ring with identity and D is an integral domain with quotient field K. If R has total quotient ring T, then by an overring S of R we mean a ring S such that $R \subseteq S \subseteq T$. The set of natural numbers will be denoted by ω and ω_o is the set of nonnegative integers. If A is an ideal of R, then we let $A[[X]] = \{f(X) = \Sigma_{i=o}^{\infty} a_i X^i \mid a_i \in A \text{ for each } i \in \omega_o\}$ and we denote by AR[[X]] the ideal of R[[X]] which is generated by A. We denote the quotient field of D[[X]] by q.f. (D[[X]]).

2. Quotient overrings and extended ideals. Let S denote a multiplicative system in R. In studying the ideal structure of the polynomial ring R[X], one important and widely used technique is to pass to a quotient ring R_S of R and utilize the fact that $R_S[X]$ is a quotient ring of R[X] - namely, $R_S[x] = (R[X])_S$. (cf. Theorem 35.11 of [8] and Theorems 36 and 171 of [10].) One particularly important aspect of this technique is that in studying the integral domain D[X], one is able to make considerable use of the ideal structure of the Euclidean domain K[X]. For example, one can easily describe the essential valuation overrings of D[X] in terms of the essential valuation overrings of D and K[X] [4, Lemma 1]. Unfortunately, one cannot,

in general, employ such techniques in studying power series rings. In fact, it is clear from the following result, proved by Gilmer in [7], that $D[[X]]$ and $K[[X]]$ seldom share the same quotient field.

THEOREM 1. _The following conditions are equivalent._

(1) _q.f._$(D[[X]])$ = _q.f._$(K[[X]])$.

(2) $K[[X]] = (D[[X]])_{D-(0)}$.

(3) _If_ $\{a_i\}_{i=1}^{\infty}$ _is a collection of nonzero elements of_ D, _then_ $\cap_{i=1}^{\infty} a_iD \neq (0)$.

In [13, Theorem 3.8], Sheldon restates Theorem 1 in the following more general form.

THEOREM 2. If S is a multiplicative system of D, then the following statements are equivalent.

(1) _q.f._$(D_S[[X]])$ = _q.f._ $(D[[X]])$.

(2) $D_S[[X]] \subseteq (D[[X]])_{D-(0)}$.

(3) _For every sequence_ $\{s_i\}_{i=1}^{\infty}$ _of elements of_ S, $\cap_{i=1}^{\infty} s_iD \neq (0)$.

Even when the conditions of Theorem 2 hold, it need not be the case that $D_S[[X]] = (D[[X]])_S$. In fact, $D_S[[X]]$ may not be a quotient overring of $D[[X]]$. For example, if we let V denote a rank two discrete valuation ring with prime ideals $(0) \subset P \subset M$, then $V_P[[X]] \subseteq (V[[X]])_{V-(0)}$, however, $V_P[[X]]$ is not a quotient overring of $V[[X]]$. In [13], Sheldon further investigates the relationship between q.f.$(D[[X]])$ and q.f.$(D_1[[X]])$, where D_1 is an overring of D. He shows, for example, that D is completely integrally closed if and only if q.f.$(D[[X]]) \neq$ q.f.$(D_1[[X]])$ for each overring D_1 properly containing D [13, Theorem 3.4]. He also shows that if $a \in D - (0)$ and $\cap_{i=1}^{\infty} a^iD = (0)$, then q.f.$(D[[X / a]])$ has infinite transcendence degree over q.f $(D[[X]])$.

Another useful technique in considering ideals in $R[X]$ is the following:

(*) Let B be an ideal of $R[X]$ and set $A = B \cap R$. Then $\bar{B} = B/AR[X]$ is an ideal in the polynomial ring $(R/A)[X] \cong R[X]/AR[X]$. Moreover, $\bar{B} \cap (R/A) = (0)$. (cf. Theorems 28, 31 and 39 of [10].)

Even though we do have the isomorphism $(R/A)[[X]] \cong R[[X]]/A[[X]]$ for power series rings, it is not generally true that $A[[X]] = AR[[X]]$. Thus, if B is an ideal of $R[[X]]$ with $A = B \cap R$, it may happen that $B \not\supseteq A[[X]]$. The following result is given by Gilmer and Heinzer in [9, Proposition 1].

THEOREM 3. *Let* A *be an ideal of* R. *Then* $A[[X]] = AR[[X]]$ *if and only if for each countably generated ideal* $C \subseteq A$, *there exists a finitely generated ideal* B *such that* $C \subseteq B \subseteq A$.

Gilmer and Heinzer also note that the equality $A[[X]] = AR[[X]]$ need not imply that A is finitely generated. For example, if V is a valuation ring with maximal ideal M and whose set of prime ideals is of ordinal type Ω, the first uncountable ordinal, then $MV[[X]] = M[[X]]$, but M is not finitely generated [9, p. 386]. However, in order that we be able to employ globally the method described in (*), we need that $A[[X]] = AR[[X]]$ for each ideal A of R. The following result is an easy consequence of Theorem 3.

THEOREM 4. *The equality* $A[[X]] = AR[[X]]$ *holds for each ideal* A *of* R *if and only if* R *is Noetherian*.

If Q is a prime ideal of $R[[X]]$ and $P = Q \cap R$, then $Q \supseteq \sqrt{PR[[X]]}$. Thus, if we wish to apply the reduction technique of (*) only to prime ideals of $R[[X]]$, then it is sufficient to have $P[[X]] = \sqrt{PR[[X]]}$ for each prime ideal P of R. In considering radical ideals in $R[[X]]$, some results concerning nilpotent elements in $R[[X]]$ will be useful.

Let $f(X) = \Sigma_{i=o}^{\infty} a_i X^i \in R[[X]]$. If R has characteristic $n > 0$, then Fields has shown in Theorem 1 of [5] that f(X) is nilpotent if and only if there is a positive integer k such that $a_i^k = 0$ for each $i \in \omega_o$. Further, Fields gives an example to show that one cannot drop the assumption that R has finite characteristic, that is, in an arbitrary ring R it may happen that $a_i^k = 0$ for each $i \in \omega_o$, yet, f(X) is not nilpotent. Let A_f denote the ideal of R generated by the coefficients of f(X). If we impose the condition that $a^k = 0$ for each $a \in A_f$, then there exists $m \in \omega$ such that $mA_f^m = (0)$. This satisfactorily imitates the assumption of "finite characteristic" and, in fact, yields that g(X) is nilpotent for each $g(X) \in A_f[[X]]$.

More generally, we have the following result [1, Lemma 4].

LEMMA. *Let A be an ideal of R and suppose there is a positive integer k such that* $a^k = 0$ *for each* $a \in A$. *Then* $f(X)$ *is nilpotent for each* $f(X) \in A[[X]]$.

The author does not know if $f(X)$ nilpotent implies the existence of $k \in \omega$ such that $a^k = 0$ for each $a \in A_f$. Thus, in the following theorem, which is immediate from the above lemma, we are able only to give sufficient conditions on an ideal A in order that $A[[X]] \subseteq \sqrt{AR[[X]]}$.

THEOREM 5. *Let A be an ideal of R and suppose that for each countably generated ideal* $C \subseteq A$ *there exists a positive integer k and a finitely generated ideal* $B \subseteq A$ *such that* $c^k \in B$ *for each* $c \in C$. *Then* $A[[X]] \subseteq \sqrt{AR[[X]]}$.

We shall say that an ideal A is an ideal of *strong finite type* (or, an SFT-ideal) provided there exists $k \in \omega$ and a finitely generated ideal $B \subseteq A$ such that $a^k \in B$ for each $a \in A$. We say that the ring R satisfies the SFT-*property* if each ideal of R is an SFT-ideal [2]. It is easy to see that if R satisfies the SFT-property, then R satisfies the ascending chain condition for radical ideals, that is, R has Noetherian prime spectrum. The converse does not hold. For example, a rank one nondiscrete valuation ring does not satisfy the SFT-property.

Following Theorem 3, we referred to an example of a valuation ring V with maximal ideal M such that M is not finitely generated, yet $M[[X]] = MV[[X]]$. Clearly then, $M[[X]] = \sqrt{MV[[X]]}$. But M is not an SFT-ideal since it is not even the radical of a finitely generated ideal. However, if we impose globally the condition that $A[[X]] \subseteq \sqrt{AR[[X]]}$, then we obtain the following result [1, Theorem 1].

THEOREM 6. *The following conditions are equivalent.*

(1) R *satisfies the* SFT-*property*.

(2) $A[[X]] \subseteq \sqrt{AR[[X]]}$ *for each ideal* A *of* R.

(3) $P[[X]] = \sqrt{PR[[X]]}$ *for each prime ideal* P *of* R.

In view of the somewhat analogous nature of Theorems 4 and 6, it seems reasonable to ask whether the condition that $P[[X]] = PR[[X]]$ for each prime ideal

P of R implies that R is Noetherian. The author does not know the answer.

3. Krull dimension in R[[X]]. We say that R has dimension n, and write dim R = n, provided there exists a chain $P_o \subset P_1 \subset \ldots \subset P_n$ of n + 1 prime ideals of R, where $P_n \neq R$, but no such chain of n + 2 prime ideals. Seidenberg has shown in [11] that if dim R = n, then $n + 1 \leq \dim R[X] \leq 2n + 1$. It is easy to see that one also has $n + 1 \leq \dim R[[X]]$, however, the following result shows that dim R[[X]] has no upper bound [1, Theorem 1].

THEOREM 6. If R does not satisfy the SFT-property, then dim R[[X]] = ∞.

In order that a Prüfer domain D satisfy the SFT-property, it is necessary and sufficient that for each nonzero prime ideal P of D, there exists a finitely generated ideal A such that $P^2 \subseteq A \subseteq P$ [2, Proposition 3.1]. In particular, a valuation ring V satisfies the SFT-property if and only if it contains no idempotent prime ideals. A rank one nondiscrete valuation ring, therefore, provides a simple example of a Prüfer domain which does not satisfy the SFT-property. An integral domain D is called an almost Dedekind domain provided D_M is a rank one discrete valuation ring for each maximal ideal M of D. An almost Dedekind domain which is not Dedekind [8, p. 586] provides another example of a Prüfer domain which does not satisfy the SFT-property.

If D is a Prüfer domain with dim D = n, then Seidenberg has shown in [12, Theorem 4] that dim D[X] = n + 1. The analogous statement for power series rings does not hold, for, as we have just seen, D need not satisfy the SFT-property. Thus, we may have dim D[[X]] = ∞. However, if we restrict our attention to rings which satisfy the SFT-property, then we can extend Seidenberg's result to power series rings. Namely, we get the following result [2, Theorem 3.8].

THEOREM 7. Let D be a Prüfer domain with dim D = n. The following statements are equivalent.

(1) D satisfies the SFT-property

(2) Dim D[[X]] = n + 1.

(3) Dim D[[X]] < ∞.

It can be easily shown that if R is any ring with dim R = 0, then statements (1) - (3) of Theorem 7 are equivalent for R. The author knows of no example for which (1) - (3) are not equivalent. In an attempt to find such an example, one might be lead to consider rings R which satisfy the SFT-property but for which dim R[X] > 1 + dim R. To construct one such ring, let k be a field and K = k(t) a simple transcendental extension of k. Suppose that V = K + M is a rank one discrete valuation ring with maximal ideal M (e.g., take V = K[[Y]]). If we set D = k + M, then D satisfies the SFT-property and dim D = 1, yet, dim D[X] = 3 [8, Appendix 2]. In this case, however, we get that dim D[[X]] = 2 [9].

From the statement of Theorem 6, one can conclude only that there is no bound on the lengths of chains of prime ideals of R[[X]]. The proof given in [1, Theorem 1] shows, more strongly, that R[[X]] contains an infinite ascending chain of prime ideals. It is easy to see, then, that when statements (1) - (3) of Theorem 7 are equivalent for a ring R, then they are equivalent to the statement "R[[X]] satisfies the a.c.c. for prime ideals". The proof given in [1] for Theorem 6 shows, in fact, that if R does not satisfy the SFT-property, then there exists a nonzero prime ideal P of R and an infinite chain $Q_1 \subset Q_2 \subset \dots$ of prime ideals of R[[X]] such that $Q_i \cap R = P$ for each $i \in \omega$. Following a proof given by Fields in [6, Lemma 2.4], Arnold and Brewer show in [3, Proposition 1] that if Q is a prime ideal of R[[X]] such that $Q \cap R = M$ is a maximal ideal of R, then either Q = M + (X) or $Q \subseteq M[[X]]$. Thus, if D is a one dimensional integral domain which does not satisfy the SFT-property, then there exists a maximal ideal M of D and an infinite chain $Q_1 \subset Q_2 \subset \dots$ of prime ideals of D[[X]] such that $MD[[X]] \subset Q_1 \subset Q_2 \subset \dots \subset M[[X]]$.

It follows from Theorem 16.10 of [8] that if P is a prime ideal of the Prüfer domain D, then each prime ideal of D[X] contained in P[X] is the extension of a prime ideal of D. Our remarks in the preceding paragraph show that this need not occur in power series rings. We do have the following somewhat analogous result.

THEOREM 8. Let D be a finite dimensional Prüfer domain which satisfies the SFT-property, and let P be a prime ideal of D. The only prime ideals of D[[X]] contained in P[[X]] have the form $P_1[[X]]$ for some prime ideal P_1 of D.

4. Valuation overrings of D[[X]]. Let P be a prime ideal of the integral domain D. The domain $(D[X])_{P[X]}$ is a valuation ring if and only if D_P is a valuation ring. For Krull domains, the analogous statement for power series rings holds [9, Theorem 2]. In [3] Arnold and Brewer have considered the following question: If P is a prime ideal of the domain D, when is $(D[[X]])_{P[[X]]}$ a valuation ring? Necessary conditions are given in the following theorem [3, Theorem 1].

THEOREM 9. *Let P be a prime ideal of the integral domain D. If $(D[[X]])_{P[[X]]}$ is a valuation ring, then D_P is a rank one discrete valuation ring. Moreover, $(D[[X]])_{P[[X]]}$ is rank one discrete.*

The proof given in [3] for Theorem 9 shows the following: Let $(0) \subset Q \subset P$ be prime ideals of D and assume that D_P is a valuation ring with $QD_P = \cap_{n=1}^{\infty} a^n D_P$ for some $a \in P - Q$. Then $(D[[X]])_{Q[[X]]}$ is not a valuation ring. Thus, if P is a prime ideal of any valuation ring V, with dim V = n > 1, then $(V[[X]])_{P[[X]]}$ is not a valuation ring. This shows that the condition that D_P be a rank one discrete valuation ring is not sufficient to insure that $(D[[X]])_{P[[X]]}$ is a valuation ring. Our observations in the previous section yield another simple example. If D is an almost Dedekind domain which is not Dedekind, then there exists a maximal ideal M of D and an infinite chain $Q_1 \subset Q_2 \subset \ldots$ of prime ideals of D[[X]] such that $MD[[X]] \subset Q_1 \subset Q_2 \subset \ldots \subset M[[X]]$. In view of Theorem 9, $(D[[X]])_{M[[X]]}$ is not a valuation ring.

The following theorem gives sufficient conditions for $(D[[X]])_{P[[X]]}$ to be a valuation ring [3, Theorem 2].

THEOREM 10. *Suppose that P is a prime ideal of the integral domain D and that D_P is a rank one discrete valuation ring. If PD[[X]] is a prime ideal of D[[X]], then $(D[[X]])_{PD[[X]]}$ is a rank one discrete valuation ring. In particular, if PD[[X]] = P[[X]], then $(D[[X]])_{P[[X]]}$ is a valuation ring.*

While the assumption that PD[[X]] is prime is ostensibly more general than the assumption that PD[[X]] = P[[X]], the authors in [3] are unable to provide an example of a prime ideal P which satisfies the hypotheses of Theorem 10 and which has the property that $DP[[X]]$ is prime but $PD[[X]] \neq P[[X]]$. If V is a rank one non-discrete valuation ring with maximal ideal M, it is the case that $MV[[X]]$ is prime, yet, $MV[[X]] \neq M[[X]]$ (see Lemma 1 of [3] for a proof due to M. Van der Put).

The conditions given in Theorem 10 are not necessary, for one can construct a Krull domain D with a minimal prime ideal P such that $PD[[X]]$ is not prime (in particular, $PD[[X]] \neq P[[X]]$), yet, $(D[[X]])_{P[[X]]}$ is a valuation ring [3].

Although the following theorem gives necessary and sufficient conditions in order that $(D[[X]])_{P[[X]]}$ be a valuation ring, these conditions are not entirely satisfactory since they are not given in terms of the ideal structure of D[3, Theorem 3].

THEOREM 11. *Let P be a prime ideal of the integral domain D and suppose that D_P is a rank one discrete valuation ring. The following conditions are equivalent.*

(1) $(D[[X]])_{P[[X]]}$ *is a valuation ring.*

(2) $P(D[[X]])_{P[[X]]} = P[[X]](D[[X]])_{P[[X]]}$.

(3) $D_P[[X]] \cap \text{q.f.}(D[[X]]) \subseteq (D[[X]])_{P[[X]]}$.

REFERENCES

1. Arnold J., Krull dimension in power series rings, *Trans. Amer. Math. Soc.* (to appear).

2. ________, Power series rings over Prüfer domains, *Pacific J. Math.* (to appear).

3. Arnold, J. and Brewer, J., On when $(D[[X]])_{P[[X]]}$ is a valuation ring (submitted).

4. __________________, Kronecker function rings and flat D[X] - modules, *Proc. Amer. Math. Soc* 27 (1971), 483-485.

5. Fields, D., Zero divisors and nilpotent elements in power series rings, *Proc. Amer. Math. Soc.*, 27 (1971), 427-433.

6. ________, Dimension theory in power series rings, *Pacific J. Math* 35 (1970), 601-611.

7. Gilmer, R., A note on the quotient field of the domain D[[X]], Proc. Amer. Math. Soc. 18 (1967), 1138-1140.

8. __________, "Multiplicative Ideal Theory", Queen's Papers on Pure and Applied Mathematics, No. 12, Kingston, Ontario, 1968.

9. Gilmer, R. and Heinzer, W., Rings of formal power series over a Krull domain, Math. Zeit. 106 (1968), 379-387.

10. Kaplansky, I., "Commutative Rings", Allyn and Bacon, Boston, 1970.

11. Seidenberg, A., A note on the dimension theory of rings, Pacific J. Math. 3 (1953), 505-512.

12. ______________, A note on the dimension theory of rings II, Pacific J. Math. 4 (1954), 603-614.

13. Sheldon, P., How changing D[[X]] changes its quotient field, Trans. Amer. Math. Soc. 159 (1971), 223-244.

KRULL DIMENSION OF POLYNOMIAL RINGS

J.W. Brewer, P.R. Montgomery, E.A. Rutter
University of Kansas, Lawrence, Kansas

and

W.J. Heinzer
Purdue University, Lafayette, Indiana

ABSTRACT. Let Q be a prime ideal of $R[X_1, \ldots, X_n]$ and let $P = Q \cap R$. This paper investigates the relationship between the ranks of Q and $P[X_1, \ldots, X_n]$. The results are used to recover some well-known results concerning the Krull dimension of $R[X_1, \ldots, X_n]$. The paper also contains a number of examples related to questions which arose in connection with the above investigation.

Let R be a commutative unitary ring with $X_1, \ldots, X_n$ indeterminates over R. It is well-known that if R is a Noetherian ring or a Prufer domain, then $\dim (R[X_1, \ldots, X_n]) = n + \dim (R)$, where dim (S) denotes the Krull dimension of the ring S. It is customary [3] and [9] to treat these two classes of rings in a different fashion. However, Kaplansky [7], for the case of a single variable, treats the two cases in a unified manner by introducing the notion of a strong S-ring. If strong S-rings respected polynomial extensions, then this approach could be extended to several variables by means of an induction argument. But strong S-rings do not respect polynomial extension as we shall see in Section 2 and in fact, we shall show by examples that it is not possible to give an n-variable definition of strong S-ring in such a way that the approach can be extended. So, if a unified treatment is to be given for several variables, another approach must be found. The properties of strong S-rings which are crucial for proving that $\dim(R[X_1]) = 1 + \dim(R)$ are: (i) rank (P) = rank $(P[X_1])$ for each prime ideal P of R and (ii) if Q is a prime ideal of $R[X_1]$ with $Q \supset (Q \cap R)[X_1]$, then rank $(Q) = 1 +$ rank $(Q \cap R)[X_1]$. In trying to extend these ideas to n variables, one sees readily that rank (P) = rank $(P[X_1, \ldots, X_n])$ in the classical cases mentioned above and therefore the problem is to relate the rank of a prime Q of $R[X_1, \ldots, X_n]$ with that of $(Q \cap R)[X_1, \ldots, X_n]$. The principal positive result of this paper is Theorem 1, which states that for an arbitrary commutative ring R and an arbitrary prime ideal Q of $R[X_1, \ldots, X_n]$, rank (Q) = rank $((Q \cap R)[X_1, \ldots, X_n])$ + rank $(Q/(Q \cap R)[X_1, \ldots, X_n]) \leq$ rank $((Q \cap R)[X_1, \ldots, X_n]) + n$. An appealing feature of Theorem 1 is that its proof is not only brief but also elementary and therefore it seems to make available from scratch, more readily than ever before, information about the ranks of prime ideals of $R[X_1, \ldots, X_n]$. Moreover, not only can Theorem 1 be used to prove the classical dimension theorems, but it also yields immediately that $\dim(K[X_1, \ldots, X_n]) = n$, when K is a field. Some other applications include a greatly simplified proof of the "Special Chain Theorem" of Jaffard and the fact that if all maximal ideals of $R[X_1, \ldots, X_n]$ have the same rank, then R is a Hilbert ring.

Perhaps the most interesting part of the paper is Section 2 which is given over to the construction of counterexamples to questions which arise in Section 1. In particular, it is shown that strong S-rings do not respect polynomial extension and that if all maximal ideals of $R[X_1]$ have the same rank, it need not be true that all maximal ideals of R have the same rank.

Our notation is essentially that of [7]. However, we shall write "r(P)" in place of "rank (P)" for the rank of a prime ideal P of the ring R.

1. The Main Theorem and Some Applications

We proceed to the proof of our main theorem pausing only to prove a preliminary result.

LEMMA 1. Let Q be a prime ideal of $R[X_1]$ and let $P = Q \cap R$. If $Q \supset P[X_1]$, then $r(Q) = r(P[X_1]) + 1$, and for each integer $n > 1$, $r(Q[X_2, \ldots, X_n]) = r(P[X_1, \ldots, X_n]) + 1$.

Proof. Both assertions are obvious if $r(Q) = \infty$. In the finite case to prove that $r(Q) = r(P[X_1]) + 1$, we induct on r(P). If $r(P) = 0$, then $r(P[X_1]) = 0$. Since three distinct primes of $R[X_1]$ cannot have the same contraction to R, $P[X_1]$ is the unique prime ideal properly contained in Q and $r(Q) = 1 = 1 + r(P[X_1])$. Assume the result to be true for all $k < m$, where $m > 0$ and $r(P) = m$. To prove that $r(Q) = r(P[X_1]) + 1$ it suffices to prove that $r(Q_1) \leq r(P[X_1])$ for each prime ideal $Q_1 \subset Q$. Thus, we let $P_1 = Q_1 \cap R$. If $P_1 = P$, then $P[X_1] \subseteq Q_1 \subset Q$, $Q_1 = P[X_1]$ and $r(Q_1) \leq r(P[X_1])$. If $P_1 \subset P$, the induction hypothesis implies that $r(Q_1) \leq r(P_1[X_1]) + 1 \leq r(P[X_1])$.

As for the second assertion considering $R[X_1, \ldots, X_n]$ as $R[X_1][X_2, \ldots, X_n]$ we have that $R[X_2, \ldots, X_n] \cap Q[X_2, \ldots, X_n] = (R \cap Q)[X_2, \ldots, X_n]) = P[X_2, \ldots, X_n]$. Then regarding $R[X_1, \ldots, X_n]$ as $R[X_2, \ldots, X_n][X_1]$, it follows from the first part of the lemma that

$r(Q[X_2, \ldots, X_n]) = r(P[X_1, \ldots, X_n]) + 1.$

THEOREM 1. If Q is a prime ideal of $R[X_1, \ldots, X_n]$ with $Q \cap R = P$, then $r(Q) = r(P[X_1, \ldots, X_n]) + r(Q/P[X_1, \ldots, X_n]) \leq r(P[X_1, \ldots, X_n]) + n.$

Proof. We use induction on n, the case $n = 1$ being immediate from Lemma 1. Therefore, we assume the result for all $k < n$ and set $Q_1 = Q \cap R[X_1]$. If $Q_1 = P[X_1]$, the result is immediate from the induction hypothesis upon regarding $R[X_1, \ldots, X_n]$ as $R[X_1][X_2, \ldots, X_n]$. If $P[X_1] \subset Q_1$, then Lemma 1 implies that $r(Q_1[X_2, \ldots, X_n]) = 1 + r(P[X_1, \ldots, X_n])$. Moreover, the induction assumption implies that $r(Q) = r(Q_1[X_2, \ldots, X_n]) + r(Q/Q_1[X_2, \ldots, X_n]) \leq (n-1) + r(Q_1[X_2, \ldots, X_n])$. Combining these results we have that $r(Q) = 1 + r(P[X_1, \ldots, X_n]) + r(Q/Q_1[X_2, \ldots, X_n]) \leq r(P[X_1, \ldots, X_n]) + n$.
Since $1 + r(Q/Q_1[X_2, \ldots, X_n]) \leq r(Q/P[X_1, \ldots, X_n])$, $r(Q) \leq r(P[X_1, \ldots, X_n]) + r(Q/P[X_1, \ldots, X_n])$. But the reverse of this last inequality is clear, so $r(Q) = r(P[X_1, \ldots, X_n]) + r(Q/P[X_1, \ldots, X_n])$.

We remark that Theorem 1 is of no greater depth than the fact that there cannot exist in $R[X_1]$ a chain of three distinct prime ideals having the same contraction to R. This fact in turn is really no deeper than the fact that $K[X_1]$ is a PID if K is a field. Thus, we obtain an elementary proof of the next result. The customary approach is to make use of the fact that fields are Noetherian rings and then appeal to the Hilbert Basis Theorem and to Krull's Principal Ideal Theorem.

COROLLARY 1. If K is a field, then $\dim(K[X_1, \ldots, X_n]) = n$.

Proof. The chain $(0) \subset (X_1) \subset \ldots \subset (X_1, X_2, \ldots, X_n)$ shows that $\dim(K[X_1, \ldots, X_n]) \geqslant n$ and that $\dim(K[X_1, \ldots, X_n]) \leqslant n$ is clear from Theorem 1.

In order to illustrate how Theorem 1 can be used to prove the other classical dimension theory results, we generalize a little to bring the argument into focus.

COROLLARY 2. Let $\mathcal{S}$ be a class of rings which is closed under localizations. Suppose further than whenever (R,M) is a quasi-local member of $\mathcal{S}$, $r(M) = r(M[X_1, \ldots, X_n])$. Then if $S \in \mathcal{S}$, $\dim(S[X_1, \ldots, X_n]) = n + \dim(S)$. In particular, if R is Noetherian or semi-hereditary, $\dim(R[X_1, \ldots, X_n]) = n + \dim(R)$. (Recall that the ring R is semi-hereditary if finitely generated ideals of R are projective.)

Proof. Let $R \in \mathcal{S}$ and let Q be a maximal ideal of $R[X_1, \ldots, X_n]$. If $P = Q \cap R$, then since $r(Q) = r(Q(R[X_1, \ldots, X_n])_{R \setminus P})$ and since $\mathcal{S}$ is closed under localizations, in order to see that $r(Q) \leqslant n + \dim(R)$, it suffices to treat the quasi-local case (R,P) with $Q \cap R = P$. By Theorem 1, $r(Q) \leqslant n + r(P[X_1, \ldots, X_n]) = n + r(P) = n + \dim(R)$.

If R is Noetherian and if P is a prime ideal of R, then $r(P) = r(P[X_1, \ldots, X_n])$ by the Principal Ideal Theorem [7, p. 110] and its converse [7, p. 112].

If R is semi-hereditary and if P is a prime ideal of R, then R_P is a valuation domain [2, p. 113]. Thus, we need only see that if V is a valuation domain having maximal ideal P, then $r(P) = r(P[X_1, \ldots, X_n])$. We prove that if Q is a nonzero prime ideal of $V[X_1, \ldots, X_n]$ with $Q \subset P[X_1, \ldots, X_n]$, then $Q = (Q \cap V)[X_1, \ldots, X_n]$. It clearly suffices to prove that $Q \subseteq (Q \cap V)[X_1, \ldots, X_n]$. If $f \in Q$, $f \neq 0$, then $f = ag$ where $a \in V$ and some coefficient of g is a unit. Then $g \notin P[X_1, \ldots, X_n] \supset Q$ and thus $a \in Q \cap V$. We conclude that $f = ag \in (Q \cap V)[X_1, \ldots, X_n]$. It now follows that the prime ideals of $V[X_1, \ldots, X_n]$ contained in $P[X_1, \ldots, X_n]$ are extended from V and that $r(P[X_1, \ldots, X_n]) = r(P)$.

Kaplansky called an integral domain an S-domain [7, p. 26] if for each rank one prime ideal P of R, $P[X_1]$ is a rank one prime ideal of $R[X_1]$. A ring R is then called a strong S-ring if for each prime ideal N of R, R/N is an S-domain. This latter condition is easily seen to be equivalent to the requirement that adjacent primes of R extend to adjacent primes of $R[X_1]$. The class of strong S-rings is then clearly closed under localizations and homomorphic images. Since in computing the ranks of extended primes of $R[X_1]$, there is no loss of generality in first localizing, the condition that R be a strong S-ring is equivalent to the requirement that $\dim (D[X_1]) = 2$ for each one-dimensional quasi-local domain D which is the homomorphic image of a localization of R. This in turn is the same as requiring that the integral closure of each one-dimensional quasi-local domain D which is the homomorphic image of a localization of R be a Prüfer domain [8, p. 511]. Moreover, Kaplansky proved that if R is a strong S-ring, $r(P) = r(P[X_1])$ for each prime ideal P of R and it follows easily from this that $\dim (R[X_1]) = 1 + \dim (R)$.

If C is a class of rings closed under localizations and quotients

by prime ideals and having the additional property that for each R ,
dim $(R[X_1]) = 1 + \dim (R)$, it is clear that dim $(D[X_1]) = 2$ for each one-dimensional quasi-local domain in C. Thus, the strong S-rings form the largest class of rings having all three of these properties. However, this is a strictly smaller class than the class of all rings satisfying the "dimension formula" for we show in Example 5 of Section 2 that there exists a domain D which is not a strong S-ring but is such that $r(P[X_1, \ldots, X_n]) = r(P)$ for each prime P of D and each positive integer n. It then follows from Theorem 1 that dim $(D[X_1, \ldots, X_n]) = n + \dim (D)$.

Using the fact that the integral closure of a one-dimensional quasi-local S-domain is a Prüfer domain, it can be shown that if P is a rank one prime ideal of a strong S-ring R, then $P[X_1, \ldots, X_n]$ is a rank one prime ideal of $R[X_1, \ldots, X_n]$ for each positive integer n. In view of this one might hope to show that $r(P[X_1, \ldots, X_n]) = r(P)$ for any prime P of a strong S-ring. It would then follow from Theorem 1 that dim $(R[X_1, \ldots, X_n]) = n + \dim (R)$. We shall show, however, that this is not the case. In fact, we exhibit in Example 3 of Section 2 a strong S-ring R such that dim $(R[X_1, X_2]) > 2 + \dim (R)$.

As Robert Gilmer pointed out to us, the first assertion of Theorem 1 can be deduced from the "Special Chain Theorem" of Jaffard [5, p. 35]. In fact, the two results are equivalent.

A chain $C = \{Q_0 \subset Q_1 \subset \ldots \subset Q_m\}$ of prime ideals of $R[X_1, \ldots, X_n]$ is called a <u>special chain</u> if for each Q_i, the ideal $(Q_i \cap R)[X_1, \ldots, X_n]$ belongs to C. With this notation we prove

COROLLARY 3. (Jaffard) If Q is a prime ideal of $R[X_1, \ldots, X_n]$ of finite rank, then $r(Q)$ can be realized as the length of a special chain of primes of $R[X_1, \ldots, X_n]$ with terminal element Q. In particular, if R is finite dimensional, then $\dim (R[X_1, \ldots, X_n])$ can be realized as the length of a special chain of primes of $R[X_1, \ldots, X_n]$.

Proof. The second assertion is immediate from the first and for the proof of the first, use Theorem 1 and induction on $r(Q)$.

If Q is a prime ideal of $R[X_1]$ with $r(Q) < \infty$, then by Corollary 3 there exists a special chain in $R[X_1]$ terminating in Q and having length $r(Q)$. In fact, it is not hard to show that there exists a special chain $C = \{Q_0 \subset Q_1 \subset \ldots \subset Q\}$ of length $r(Q)$ and such that the corresponding chain $(Q_0 \cap R) \subseteq Q_1 \cap R \subseteq \ldots \subseteq Q \cap R$ of contracted ideals is saturated--that is, between two distinct members of the chain no prime ideal can be inserted. This fact is not true for two variables as we shall see in Example 3 of Section 2. Furthermore, even for one variable, we shall see in Example 4 that it is not always possible to find a special chain of length $r(Q)$ such that the corresponding chain of contracted ideals has length $r(Q \cap R)$.

Our next result is in the spirit of Exercises 3 and 4, p. 114 of [7]. See also p. 126 of [6].

COROLLARY 4. (a) If all maximal ideals of $R[X_1, \ldots, X_n]$ have the same rank, then R is a Hilbert ring.

(b) Suppose that R is a Hilbert ring and that all maximal ideals of R have the same rank. If n is a positive integer such that $r(M) = r(M[X_1, \ldots, X_n])$ for each maximal ideal M of R, then all maximal ideals of $R[X_1, \ldots, X_n]$ have the same rank. In particular, if R is either Noetherian or semi-hereditary, then all maximal ideals of $R[X_1, \ldots, X_n]$ have the same rank.

Proof. (a): Since R is a Hilbert ring if and only if $R[X_1, \ldots, X_m]$ is a Hilbert ring for each positive integer m, it suffices to prove (a) for $n = 1$. Thus, let M' be a maximal ideal of $R[X_1]$ and set $M = M' \cap R$. If M is not maximal, say $M \subset P$, then $r(M') = 1 + r(M[X_1]) < 1 + r(P[X_1])$ and $P[X_1]$ is not a maximal ideal. It follows that M is maximal and that R is a Hilbert ring.

(b): We first make a couple of easy observations. If D is an integral domain, then there is a one-one correspondence between prime ideals P' of $D[X_1, \ldots, X_m]$ with $P' \cap D = (0)$ and all prime ideals of $K[X_1, \ldots, X_m]$, K the quotient field of D. Moreover, since all maximal ideals of $K[X_1, \ldots, X_m]$ have rank m, (a well-known fact easily proved by induction using Corollary 1 and elementary properties of Hilbert rings) if P' is a maximal ideal of $D[X_1, \ldots, X_m]$ such that $P' \cap D = (0)$, then $r(P') = m$. To prove (b), let M'_1 and M'_2 be maximal ideals of $R[X_1, \ldots, X_n]$ and set $M_i = M'_i \cap R$. By forming $R[X_1, \ldots, X_n]/(M_i[X_1, \ldots, X_n])$ and applying the above observation, we see that $r(M'_i/(M_i[X_1, \ldots, X_n])) = n$. Thus, $r(M'_1) = r(M_1[X_1, \ldots, X_n]) + n = r(M_1) + n = r(M_2) + n = r(M_2[X_1, \ldots, X_n]) + n = r(M'_2)$.

The first two examples of Section 2 show that neither part of Corollary 4 can be strengthened. The first of these examples is an integrally closed Hilbert domain D of dimension one such that not all maximal ideals of $D[X_1]$ have the same rank. The second is an integrally closed Hilbert domain D such that all maximal ideals of $D[X_1]$ have the same rank but such that this is not true of D.

2. Examples

We present here the examples promised in the preceding section. The examples are rather technical but very illuminating and the reader wishing to gain insight into the relationship between ranks of primes of R and ranks of primes of $R[X_1, \ldots, X_n]$ would profit from studying them.

Despite the fact that our first construction is used to give the examples of least importance to this paper, it does possess a certain generality and for its possible usefulness in other contexts, we give the construction in that generality.

Let R be a Prüfer domain with quotient field K. We would like to find a domain D whose local behavior is identical to that of R except at a single prime ideal P of D and at P we wish certain pathology--namely, we want D_P to be a one-dimensional, quasi-local, integrally closed domain which is not a valuation domain. This we are able to do as we shall see from Theorem 2. We require some preliminaries.

If S is a commutative ring and if $Y = \{Y_\alpha\}$ is a collection of indeterminates over S, recall that S(Y) denotes the ring $(S[Y])_T$, where T is the set of all $f \in S[Y]$ such that c(f), the ideal of S generated by the coefficients of f, is equal to S. See [3, p. 379] for the pertinent facts about S(Y).

PROPOSITION 1. Let R be a Prüfer domain, $X = \{X_\alpha\}$ a collection of indeterminates over R and let P be a prime ideal of R.

1) If Q' is a prime ideal of R[X] with $Q' \subseteq P[X]$, then

$Q' = (Q' \cap R)[X]$.

2) $Q \to QR(X)$ is a one-one lattice preserving correspondence between the set of all prime ideals of R and the set of all prime ideals of R(X). Moreover, $R_Q(X) = (R(X))_{QR(X)} = (R[X])_{Q[X]}$.

3) If R is a Hilbert ring, then R(X) is a Hilbert ring.

Proof. 1): There exists a one-one correspondence between all prime ideals of R[X] contained in P[X] and all prime ideals of $(R[X])_{P[X]} \supseteq (R[X])_{R \setminus P} = R_P[X]$. Thus, $(R[X])_{P[X]}$ is a ring of quotients of $R_P[X]$ and, in fact, $(R[X])_{P[X]} = (R_P[X])_{PR_P[X]}$. Thus, we may start over with R a valuation domain and P its maximal ideal. If $Q' \subseteq P[X]$ and if $f \in Q'$, then $c(f) = (a_1, \ldots, a_n) = a(a'_1, \ldots, a'_n)$ where $a'_i = 1$ for some i. Thus, $f = ag$ where $g \notin P[X] \supseteq Q'$ and it follows that $a \in Q' \cap R$. Hence, $f \in (Q' \cap R)[X]$ and $Q' \subseteq (Q' \cap R)[X]$.

2): That $Q \to QR(X)$ is a one-one correspondence between the set of all prime ideals of R and the set of all prime ideals of R(X) now follows from 1) and [3, p. 380]. As for the second assertion of 2), if $S = \{f \in R[X] \mid c(f) = R\}$, then $Q[X] \cap S = \emptyset$ implies that $(R[X])_{Q[X]} = (R(X))_{QR(X)}$. On the other hand, $R_Q(X) = (R_Q[X])_T$, where $T = \{g \in R_Q[X] \mid c(g) = R_Q\} = R_Q[X] \setminus QR_Q[X]$. Therefore, $R_Q(X) = (R[X])_{Q[X]}$.

3): We need only prove that for each non-maximal prime ideal QR(X) of R(X), $QR(X) = \bigcap_H HR(X)$, where $\{HR(X)\}$ is the set of all prime ideals of R(X) properly containing QR(X). $HR(X) \supset QR(X)$ if and only if $H \supset Q$ if and only if $H[X] \supset Q[X]$. Thus, $(\bigcap_H HR(X)) \cap R[X] = \bigcap_H H[X] = Q[X] = QR(X) \cap R[X]$. Therefore, since $\bigcap_H HR(X)$ and QR(X) are extended ideals of R(X) having the same contraction to R[X], they are equal.

We remark that in the above proposition, the valuation domains R_Q and $R_Q(X)$ have the same value group and consequently, they have identical prime ideal structures.

THEOREM 2. Let R be a Prüfer domain with quotient field K and let Y_1, Y_2 be indeterminates over K. Consider $T = R(Y_1, Y_2)$, $U = (K[Y_1, Y_2])_{(Y_1)} = K(Y_2) + M$, and $V = K + M$. Put $D = T \cap V$ and $P = M \cap D$. Then

1) The quotient field of D is $K(Y_1, Y_2)$.

2) $D_P = V$ is one-dimensional, quasi-local, integrally closed and not a valuation domain.

3) P is a maximal ideal of D.

4) D is integrally closed.

5) $D[1/Y_1] = T$.

6) $Q \to QT$ is a one-one correspondence between all prime ideals of D distinct from P and all prime ideals of T. Moreover, $D_Q = T_{QT} = R_{(QT \cap R)}(Y_1, Y_2)$.

7) If R is a Hilbert ring, then D is a Hilbert ring.

<u>Proof</u>. 1): $R[Y_1, Y_1Y_2] \subseteq D$.

2): If S is an arbitrary multiplicative system in D, then $D_S = T_S \cap V_S$. In particular, if $S = D \setminus P$, then $V_S = V$. By Proposition 1, 2), since $R\setminus(0) \subseteq K\setminus(0) \subseteq S$, each nonzero prime ideal of $T = R(Y_1, Y_2)$ meets S and it follows that $T_S = K(Y_1, Y_2)$. For the second assertion, see [3, p. 560].

3): Let $d \in D\setminus P$. Then $d = k + m$, $k \in K\setminus(0)$, $m \in M$. Write $k = a/b$, $a, b \in R\setminus(0)$. Since $R \subseteq D$, $bd = a + bm \in dD$. Thus, $bm = bd - a \in M \cap D = P$ and hence, $a = bd - bm \in dD + P$.

$a + Y_1 \in dD + P$ and $a + Y_1$ is a unit in V since $a \in R\setminus(0)$. But $a + Y_1$ is a unit in T by definition. It follows that $dD + P = D$.

4): T and V are integrally closed.

5): $D[1/Y_1] = T[1/Y_1] \cap V[1/Y_1]$ and $V[1/Y_1] = K(Y_1, Y_2)$.

6): By 5), we need only see that P is the lone prime ideal of D which contains Y_1. Suppose not and let Q be a prime ideal of D containing Y_1 and distinct from P. Since P is maximal, $P \cap (D\setminus Q) \neq \emptyset$ and $D_Q = T_{D\setminus Q} \cap V_{D\setminus Q} = T_{D\setminus Q}$. Y_1 is a unit in T and hence in $T_{D\setminus Q} = D_Q$. But $Y_1 \in QD_Q$. That $T_{QT} = R_{(QT \cap R)}(Y_1, Y_2)$ follows from Proposition 1.

7): Since P is maximal, if Q is a non-maximal prime ideal of D, then QT is a non-maximal prime ideal of T. Thus, by Proposition 1, 3), QT is the intersection of the prime ideals of T properly containing it. It now follows easily that D is a Hilbert ring.

Using Theorem 2, we can now show that neither part of Corollary 4 can be strengthened. Throughout, Z denotes an indeterminate over the integral domain D.

EXAMPLE 1. <u>This is an example of an integrally closed Hilbert domain</u> D <u>of dimension one having the property that</u> $\dim(D[Z]) = 3$. <u>Moreover, some maximal ideals of</u> D[Z] <u>have rank two and some have rank three</u>.

Let R be a Dedekind domain and let K denote the quotient field of R. The notation being as in Theorem 2, since $\dim(D[Z]) \leq 3$, it suffices to prove the "moreover" assertion. If Q is a prime ideal of D distinct from P, D_Q is a DVR and rank $(Q[Z]) = 1$. Thus, maximal ideals of D[Z] contracting to Q have rank two. Since D is integrally closed and D_P is not a valuation domain, rank (P[Z]) = 2 [3, p. 218]. A maximal ideal of D[Z] contracting to P must have rank three.

EXAMPLE 2. This is an example of an integrally closed Hilbert domain D with the following two properties: (1) dim (D) = 2, but D has a maximal ideal of rank one. (2) dim (D[Z]) = 3 and each maximal ideal of D[Z] has rank three. Our problem is to choose the appropriate R and then apply Theorem 2. Moreover, since the D constructed will be a Hilbert ring, a maximal ideal of D[Z] will contract to D in a maximal ideal of D, and because the rank of a maximal ideal of D[Z] is one more than the rank of its contraction, we will be finished as soon as each maximal ideal of R has rank two. Of course, we are using here the fact that rank (P[Z]) = 2. To choose such an R, let R be a Bezout domain having divisibility group H as given in [4, p. 1371] and such that H has infinitely many prime filters. (The terminology of [4] is "minimal primes", but "prime filters" seems to be more standard.) This R will suffice.

EXAMPLE 3. We are going to construct a two-dimensional quasi-local domain (R, M) having the following properties:

(1) R has a unique rank one prime ideal P,

(2) P has the property that both R_P and R/P are DVR's, and

(3) R has valuative dimension three. (Recall that the integral domain D has valuative dimension $m < \infty$ if there exists a rank m valuation domain between D and its quotient field but no valuation domain of larger rank.)

Supposing for the moment that we have done this, let's consider the implications.

R is a strong S-ring. To see this, notice that $P[X_1]$ has rank one since R_P is a DVR and that $P[X_1] \subset M[X_1]$ are adjacent since R/P is a DVR.

$R[X_1]$ is not a strong S-ring. This is a consequence of the following theorem of Arnold [1, p. 323]: Let D be an integral domain of finite valuative dimension k. If m is a positive integer such that $m \geqslant k-1$, then $\dim (D[X_1, \ldots, X_m]) = m + k$. In our case, $k = 3$ and so $\dim (R[X_1, X_2]) = 5 > 1 + \dim (R[X_1])$.

If Q is a rank five maximal ideal of $R[X_1, X_2]$, then no special chain of length five with terminal element Q can contain $P[X_1, X_2]$. Thus, in contrast to the single variable case, for two variables it need not be true that all ranks of prime ideals are realizable as the lengths of special chains where the contracted chain is saturated. This assertion follows again from the fact that R_P and R/P are DVR's.

Now to the construction. In the hope that some clarity may be achieved by casting the example in a more general setting, we isolate the following portion of the argument.

LEMMA 2. Let (V, M) and W, N) be quasi-local domains containing a common subfield K and set $R = K + (M \cap N)$. Then R is a quasi-local domain with maximal ideal $M \cap N$. Moreover, if W is a rank one valuation domain and if V and W have the same quotient field, then for each nonzero nonmaximal prime ideal Q of R, $R_Q = V_{Q'}$, for some prime ideal Q' of V.

Proof. That R is a domain and that $M \cap N$ is an ideal of R consisting entirely of nonunits is obvious. If $r = a + m \in R \setminus (M \cap N)$, with $a \in K$, $m \in (M \setminus N)$ then $a \neq 0$ and so

$1/r = (1/a) + (-m)/(a(a + m)) \in R$ since $(-m)/(a(a + m)) \in M \cap N$.

To prove the second assertion, suppose that Q is a nonzero prime ideal of R with $Q \subset (M \cap N)$ and choose $t \in (M \cap N) \setminus Q$. If $s \in V$, then since W has rank one, there exists a positive integer c such that $t^c s \in N$. Therefore, $t^c s \in M \cap N$ and $s \in R_Q$. Since t is a unit in R_Q but not in V, $V \subset R_Q$.

So, to construct the promised R, we begin by building valuation domains V and W so as to enable us to apply Lemma 2.

Let K be a field with Y_1, Y_2 and Y_3 indeterminates over K. Let V_1 be the valuation domain of the Y_3-adic valuation on $K(Y_1, Y_2)[Y_3]$. V_1 is a DVR with residue field $K(Y_1, Y_2)$. On $K(Y_1, Y_2)$, build a rank one valuation domain V^* by mapping Y_1 and Y_2 to two rationally independent real numbers, say $Y_1 \to 1$ and $Y_2 \to \pi$. If f denotes the canonical map from V_1 onto $K(Y_1, Y_2)$, then $f^{-1}(V^*) = V$ is a rank two valuation domain of the form $K + M_1$. Moreover, if P_1 is the rank one prime of V, $V_1 = V_{P_1}$. (We refer the reader to [3, p. 190] for a more general treatment of this "pull-back" construction of valuation domains.) Now, if W is the valuation domain of the $(Y_3 + 1)$-adic valuation on $K(Y_1, Y_2)[Y_3]$, then W is a DVR of the form $K(Y_1, Y_2) + M_2$ and $W \neq V_{P_1}$. Set $R = K + M_1 \cap M_2$. By Nagata's Theorem [3, p. 262], $D = V \cap W$ is a two-dimensional Prüfer domain with quotient field $K(Y_1, Y_2, Y_3)$ and the maximal ideals of D are $M_1 \cap D$ and $M_2 \cap D$. Thus $M_1 \cap M_2$ is a nonzero ideal of D and it follows that the quotient field of $M_1 \cap M_2$ is $K(Y_1, Y_2, Y_3)$. From this we see also that the quotient field of R is $K(Y_1, Y_2, Y_3)$. Since $R \subseteq K + M_2$ and since the valuative dimension of $K + M_2$ is three [3, p. 561], the valuative dimension of R

is three. Moreoever, by Lemma 2, R is quasi-local with maximal ideal $M_1 \cap M_2$. But $P = P_1 \cap R$ is a nonzero nonmaximal prime of R for $P \neq (0)$ and if $P \supseteq M_1 \cap M_2$, then $P_1 \supseteq (M_1 \cap D)(M_2 \cap D)$ which is impossible. So again by Lemma 2, P is the unique rank one prime of R and $R_P = V_1$, a DVR. It remains only to show that R/P is a DVR. It suffices to show that $R/P = D/(P_1 \cap D)$. Now $R/P \subseteq D/(P_1 \cap D)$ and since for $d \in D$ there exists $a \in K$ with $d - a \in M_1$, to show that $D/(P_1 \cap D) \subseteq R/P$, it suffices to show that for $d \in M_1 \cap D$, there exists $b \in R$ such that $\overline{d} \in D/(P_1 \cap D)$ equals $\overline{b} \in R/P$. But $P_1 \cap D$ and $M_2 \cap D$ are comaximal and so such a b does exist.

As a variant on the preceding construction we have

EXAMPLE 4. *This is an example of a domain* R *such that there exists a prime ideal* Q *of* $R[X_1]$ *having the property that for each special chain of primes of* $R[X_1]$ *terminating in* Q *and having length* $r(Q)$, *the corresponding chain of contracted primes fails to have length* $r(Q \cap R)$.

It will suffice to find a quasi-local domain (R, M) having the following properties:

(1) The prime ideal structure of R is given by the diagram

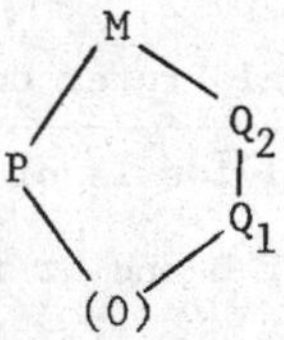

(2) $\dim (R [X_1]) = 5$.

(3) $(0) \subset Q_1[X_1] \subset Q_2[X_1] \subset M[X_1]$ is a saturated chain.

To build such a domain, let K be a field with Y_1, Y_2, Y_3 and Y_4 indeterminates over K. Consider $(K(Y_1, Y_2, Y_3)[Y_4])_{(Y_4)} = K(Y_1, Y_2, Y_3) + M_1$ and $(K(Y_1)[Y_2])_{(Y_2)} = K(Y_1) + M_2$. Let $M_V = M_2 + M_1$ and set $V = K + M_V$. From Lemma 2 and standard facts about the "D + M-construction", V is a two-dimensional quasi-local domain with maximal ideal M_V and with M_1 as unique rank one prime. Note that $V[X_1]$ has dimension five with primes properly between (0) and $M_1[X_1]$ and between $M_1[X_1]$ and $M_V[X_1]$.

Now consider $(K(Y_1, Y_2, Y_3)[Y_4])_{(Y_4+1)} = K(Y_1, Y_2, Y_3) + N_1$, $(K(Y_1, Y_2)[Y_3])_{(Y_3+1)} = K(Y_1, Y_2) + N_2$ and define a rank one valuation domain on $K(Y_1, Y_2)$ of the form $K + N_3$ by giving Y_1 and Y_2 values 1 and π. Let $W = K + N$ where $N = N_3 + N_2 + N_1$. Then W is a rank three valuation domain of $K(Y_1, Y_2, Y_3, Y_4)$.

Let $M = M_V \cap N$ and $R = K + M$. Note that $Y_1Y_2, Y_2, Y_3Y_4, Y_4 \in M$ so R has quotient field $K(Y_1, Y_2, Y_3, Y_4)$. Moreover, for each nonzero $a \in M$, $R[1/a] = V[1/a] \cap W[1/a]$. Thus, if $a = Y_2(Y_4 + 1)$, then $a \in M$ and $V[1/a] = V_{M_1} = K(Y_1, Y_2) + M_1$ and $W[1/a] = K(Y_1, Y_2, Y_3, Y_4)$. So, letting $P = M_1 \cap R$, we have $R_P = V_{M_1} = K(Y_1, Y_2) + M_1$.

Taking $a = Y_4Y_2 \in M$, $V[1/a] = K(Y_1, Y_2, Y_3, Y_4)$ while $W[1/a] = W_{(N_2+N_1)} = K(Y_1, Y_2) + (N_2 + N_1)$. Taking $a = Y_4(Y_3 + 1) \in M$, $V[1/a] = K(Y_1, Y_2, Y_3, Y_4)$ while $W[1/a] = W_{N_1} = K(Y_1, Y_2, Y_3) + N_1$. It follows that if $Q_1 = N_1 \cap R$ and $Q_2 = (N_2 + N_1) \cap R$, then $R_{Q_1} = W_{N_1}$ and $R_{Q_2} = W_{(N_2+N_1)}$.

This R has the desired properties.

EXAMPLE 5. *This is an integral domain* R *which is not a strong S-ring but which does have the property that for each prime* P *of* R *and each positive integer* n, $r(P) = r(P[X_1, \ldots, X_n])$. We are indebted to J.T. Arnold for the idea.

Let K be a field with Z_1, Z_2 indeterminates over K. Let $V = K(Z_1) + M$ be the valuation domain of the Z_2-adic valuation on $K(Z_1, Z_2)$ and set $D = K + M$. D is a one-dimensional integrally closed quasi-local domain of valuative dimension two. Thus, by Arnold's Theorem [1, p. 323], $\dim (D[X_1, \ldots, X_m]) = m + 2$ for each positive integer m. Put $R = D[X_1]$ and let P be a prime ideal of R. To show that $r(P) = r(P[X_1, \ldots, X_n])$, we merely handle the few cases which arise. If $P = M[X_1]$, then $r(P) = 2$ since R is integrally closed but not a valuation domain. Thus, $r(M[X_1, Y_1, \ldots, Y_n]) \geqslant 2$. But $\dim (D[X_1, Y_1, \ldots, Y_n]/M[X_1, Y_1, \ldots, Y_n]) = \dim ((D/M)[X_1, \ldots, Y_n]) = n + 1$ and $\dim (D[X_1, Y_1, \ldots, Y_n]) = n + 3$. Hence, we must have that $r(M[X_1, Y_1, \ldots, Y_n]) = 2$. If $P \cap D = (0)$, then R_P is a DVR and $r(P) = 1 = r(P[Y_1, \ldots, Y_n])$. If $P \cap D \neq (0)$ and if $P \neq M[X_1]$, then $r(P) = 3$. By Lemma 1, $r(P[Y_1, \ldots, Y_n]) = r(M[X_1, Y_1, \ldots, Y_n]) + 1 = 3$.

REFERENCES

[1] Arnold, J.T., On the dimension theory of overrings of an integral domain, *Trans. Amer. Math. Soc.* 138 (1969), 313-326.

[2] Endo, S., On semi-hereditary rings, *J. Math. Soc. Jap.* 13 (1961), 109-119.

[3] Gilmer, R.W., *Multiplicative Ideal Theory*, Queen's Papers on Pure and Applied Mathematics, Kingston, Ontario, 1968.

[4] Heinzer, W. J., J-Noetherian integral domains with 1 in the stable range, _Proc_. _Amer_. _Math_. _Soc_. 19 (1968), 1369-1372.

[5] Jaffard, P., _Theorie de la Dimension dans les Anneaux de Polynomes_, Gauthier-Villars, Paris, 1960.

[6] Kaplansky, I., _Commutative Rings_, Queen Mary College Mathematics Notes, London, 1966.

[7] __________, _Commutative Rings_, Allyn and Bacon, Boston, 1971.

[8] Seidenberg, A., A note on the dimension theory of rings, _Pac_. _J_. _Math_. 3 (1953), 505-512.

[9] __________, _On the dimension theory of rings_, Pac. J. Math. 4 (1954), 603-614.

A NOTE ON THE FAITHFULNESS OF THE FUNCTOR $S \otimes_R (-)$

J.W. Brewer and E.A. Rutter

University of Kansas, Lawrence, Kansas

ABSTRACT. If R and S are rings with S a unitary extension of R, the faithfulness of the functor $S \otimes_R (-)$ is studied.

Let R and S be unitary rings with S an extension of R. Since the functor $S \otimes_R (-)$ is additive, it is faithful if and only if it reflects zero maps [2, pp. 52 and 56]. It is natural to ask whether or not this is equivalent to the apparently weaker condition that $S \otimes_R (-)$ reflects zero objects. Of course, this is the case if S is a flat extension of R [2, Prop. 7.2, p. 57]. The purpose of this note is to show that in general, however, this is not so. In fact, we prove that $S \otimes_R (-)$ is faithful if and only if S is a pure extension of R in the sense of Cohn [1] and we give an example of integral domains R and S having the following two properties:

(1) S is not a pure extension of R and

(2) $S \otimes_R (-)$ reflects zero objects.

Let $0 \longrightarrow E \longrightarrow F$ be an exact sequence of right R-modules. The sequence is said to be pure if and only if the sequence $0 \longrightarrow E \otimes_R M \longrightarrow F \otimes_R M$ is exact for each R-module M. When R is a subring of S and the sequence $0 \to R \to S$ is pure, we say S is a <u>pure extension</u> of R.

Theorem. Let $R \subseteq S$ be rings, R a subring of S. In order that the functor $S \otimes_R (-)$ be faithful, it is necessary and sufficient that S be a pure extension of R.

<u>Proof</u>. Since the functor $S \otimes_R (-)$ is additive, it is faithful if and only if it reflects the zero mapping; that is, if and only if for each nonzero R-homomorphism $f : M' \to M$ of R-modules, the mapping $1_S \otimes f : S \otimes_R M' \to S \otimes_R M$ is nonzero. Assuming that S is a pure extension of R, let $f : M' \to M$ be a nonzero R-homomorphism and consider the following commutative diagram.

$$\begin{array}{ccc} S \otimes_R M' & \xrightarrow{1_S \otimes f} & S \otimes_R M \\ \uparrow & & \uparrow \\ R \otimes_R M' & \xrightarrow{1_R \otimes f} & R \otimes_R M \end{array}$$

Since the vertical arrows are monomorphisms and since $1_R \otimes f$ is nonzero, $1_S \otimes f$ is nonzero and it follows that $S \otimes_R (-)$ is faithful.

Suppose now that S is not a pure extension of R and let M be an R-module such that $0 \to R \otimes_R M \to S \otimes_R M$ is not exact. Identifying $R \otimes_R M$ with M, we see that this is equivalent to the condition that the canonical map $e_M : M \to S \otimes_R M$ is not a monomorphism. Let $M' = \ker e_M$ and denote by f the injection of M' into M. The diagram

$$\begin{array}{ccc} S \otimes_R M' & \xrightarrow{1_S \otimes f} & S \otimes_R M \\ \uparrow e_{M'} & & \uparrow e_M \\ M' & \xrightarrow{f} & M \end{array}$$

commutes and by the definition of f, $e_M \circ f = 0$. But $e_{M'}(M')$ generates

$S \otimes_R M'$ as an S-module and therefore $1_S \otimes f = 0$. We conclude that $S \otimes_R (-)$ is not faithful.

Before providing the example promised in the introduction, we make some preliminary observations. Let $R \subseteq S$ be unitary rings with R a subring of S. Theorem 2.4 of [1] shows that S is a pure extension of R if and only if each system of linear equations with coefficients in R which is solvable in S is solvable in R. Thus, if S is a pure extension of R, then each linear equation which is solvable in S is solvable in R. We observe that this latter condition is equivalent to the condition that $AS \cap R = A$ for each right ideal A of R. To construct the desired example, let K and F be fields with F a proper subfield of K. If X is an indeterminate over K denote by S the ring of formal power series in X over K. Then $S = K + M$, where M is the maximal ideal of S, and if we set $R = F + M$, then M is a common maximal ideal of R and S. We verify first that S is not a pure extension of R. By the above observations, it suffices to find an ideal A of R such that $AS \cap R \neq A$. Let $s \in S$, $s \notin R$ and let $a \in M$, $a \neq 0$. Then $as \in R$ and if $aS \cap R = aR$, then $as = at$ for some $t \in R$ which is absurd. It follows that $aS \cap R \supset aR$. It remains to show that $S \otimes_R (-)$ reflects zero objects. Suppose N is an R-module such that $S \otimes_R N = 0$. Then $E \otimes_R N = 0$, for any S-module E since $E \otimes_R N = (E \otimes_S S) \otimes_R N = E \otimes_S (S \otimes_R N) = 0$. In particular, $S/M \otimes_R N = 0$ and $M \otimes_R N = 0$. Thus, $MN = 0$ since $M \otimes_R N$ maps onto MN via the mapping $m \otimes n \rightarrow mn$. Since $M \subseteq R$, there is a natural R-monomorphism $0 \rightarrow R/M \rightarrow S/M$. As M is a common ideal of R and S, it annihilates both R/M and S/M. Hence, these modules may be regarded in the usual way as R/M-modules. If this is done, then the R-homomorphisms and the R/M-homomorphisms between them are identical. Thus, the above monomorphism splits since R/M is a field. Therefore, the sequence $0 \to R/M \otimes_R N \to S/M \otimes_R N$ is exact and hence $R/M \otimes_R N = 0$. But $R/M \otimes_R N = N/MN$ and since $MN = 0$, we conclude that $N = 0$.

REFERENCES

1. P. Cohn, On the free product of associative rings, I. Math. Zeit. 71 (1959), 380-398.

2. B. Mitchell, Theory of Categories, New York: Academic Press, 1965.

On A PROBLEM IN LINEAR ALGEBRA

David A. Buchsbaum

and

David Eisenbud

In its vaguest and most tantalizing form, the problem referred to in the title of this paper is to say something about the solution of systems of linear equations over a polynomial ring $R = K[X_1,\dots,X_n]$, where K is a field. The difficulty is, of course, that a system of linear equations over R may not possess a set of linearly independent solutions from which all solutions may be obtained by forming linear combinations with coefficients in R.

Hilbert's Syzygy Theorem suggests a promising approach to this problem: Let φ_1 be a system of linear equations, which we will regard as a map of free R-modules, say $\varphi_1 : F_1 \longrightarrow F_o$. If we form a free resolution of the cokernel of φ_1, that is an exact sequence of free R-module of the form

$$(1)\qquad F_n \xrightarrow{\varphi_n} F_{n-1} \xrightarrow{\varphi_{n-1}} F_{n-2} \longrightarrow \cdots \longrightarrow F_2 \xrightarrow{\varphi_2} F_1 \xrightarrow{\varphi_1} F_o ,$$

then the Syzygy Theorem tells us that the system of linear equations represented by φ_{n-1} does have a full set of linearly independent solutions. More precisely, it tells us this in the graded or local cases; in general, Ker φ_{n-1} is merely stably free. Thus, replacing F_n by Ker φ_{n-1} in the sequence (1), we may replace (1) by a finite free (or stably free) resolution of the form

$$(2)\qquad 0 \longrightarrow F_n \xrightarrow{\varphi_n} F_{n-1} \longrightarrow \cdots \longrightarrow F_1 \xrightarrow{\varphi_1} F_o .$$

In order to use the approach to linear equations that Hilbert's Theorem suggests, it seems to be necessary to study the relations that must hold between the maps φ_i in a finite free resolution of the form of (2). The first result, establishing a relationship between the maps φ_i, was also obtained by Hilbert [6]. Letting $\overset{n}{\wedge}X$ denote the nth exterior power of a module X, we state it as;

<u>Theorem 1.</u> Suppose that R is a noetherian ring. If

$$(3)\qquad 0 \longrightarrow R^n \xrightarrow{\varphi_2} R^{n+1} \xrightarrow{\varphi_1} R$$

is an exact sequence of free R-modules, then there exists a non-zerodivisor $a \in R$

such that, after making the canonical identification $\overset{n}{\Lambda} R^{n^*} \approx R$ and $\overset{n}{\Lambda} R^{n+1^*} \approx R^{n+1}$, we have $\varphi_1 = a \overset{n}{\Lambda} \varphi_2{}^*$.

(Hilbert actually proved this only when R is the graded polynomial ring in two variables over a field; he did it to illustrate the application of the Hilbert function. The first proof that works in the generality in which we have stated the theorem is due to Burch [5]. The most elementary proof is contained in [7]. We shortly give a new proof of this result.)

Theorem 1 can be applied to a special case of Grothendieck's Theorem in a way that we will now describe. To do so, we will shift our point of view from polynomial rings to regular local rings; just as in the polynomial case, every module over a regular local ring has a finite free resolution.

Grothendieck's lifting problem is the following: let S be a regular local ring, and let $x \in S$ be such that $R = S/(x)$ is regular. Given a finitely generated R-module M, does there exist a finitely generated S-module $M^{\#}$ such that

(1) $M^{\#}/xM^{\#} \approx M$;

(2) x is a non-zerodivisor on $M^{\#}$?

If a module $M^{\#}$ satisfying (1) and (2) exists, it is called a lifting of M.

That the lifting problem is closely related to the structure of free resolutions is shown by the next lemma.

<u>Lemma 2.</u> Let S be any ring, and let $x \in S$ be a non-zerodivisor. Let $R = S/(x)$, and let

$$F: F_2 \xrightarrow{\varphi_2} F_1 \xrightarrow{\varphi_1} F_o$$

be an exact sequence of free R-modules. Suppose that

$$F^{\#}: F_2^{\#} \xrightarrow{\varphi_2^{\#}} F_1^{\#} \xrightarrow{\varphi_1^{\#}} F_o$$

is a sequence of free S-module which reduces modules (x) to $\mathbb{F}$. If $\varphi_1^{\#}\varphi_2^{\#} = 0$, then Coker $(\varphi_1^{\#})$ is a lifting of Coker(φ_1).

<u>Proof</u>: The hypothesis allows us to construct a short exact sequence of complexes

$$0 \longrightarrow \mathbb{F}^{\#} \xrightarrow{x} \mathbb{F}^{\#} \longrightarrow \mathbb{F} \longrightarrow 0 \quad .$$

The associated exact sequence in homology contains

$$(4) \quad H_1(\mathbb{F}) \longrightarrow H_o(\mathbb{F}^{\#}) \xrightarrow{x} H_o(\mathbb{F}^{\#}) \longrightarrow H_o(\mathbb{F}) \longrightarrow 0 \quad .$$

On the other hand, $H_o(\mathbb{F}^{\#}) = \mathrm{Coker}(\varphi_1^{\#})$, $H_o(\mathbb{F}) = \mathrm{Coker}(\varphi_1)$ and, by hypothesis $H_1(\mathbb{F}) = 0$. Thus (4) becomes

$$0 \longrightarrow \mathrm{Coker}(\varphi_1^{\#}) \xrightarrow{\cdot x} \mathrm{Coker}(\varphi_1^{\#}) \longrightarrow \mathrm{Coker}(\varphi_1) \longrightarrow 0 ,$$

and the lemma is proved.

Lemma 2 implies the liftability of various classes of modules. For example, in the situation of the lifting problem, any module M of homological dimension one can be lifted: If $0 \longrightarrow F_1 \xrightarrow{\varphi_1} F_o$ is a free presentation of M , then there exists a homomorphism $\varphi_1^{\#}$ of free S-modules which reduces to φ_1 modulo x (for example, to construct $\varphi_1^{\#}$ one can choose bases of F_1 and F_2 and lift each element of the matrix of φ_1). By lemma 2, Coker $(\varphi_1^{\#})$ is a lifting of M .

Theorem 1 can be applied to the lifting problem through the use of lemma 2. Together they imply the liftability of cyclic modules of homological dimension 2. For, if (3) is the free resolution of such a module, we may choose a map $\varphi_2^{\#}$ of free S-modules and an element $a^{\#} \in S$ which reduce, modulo (x), to φ_2 and a respectively. Letting $\varphi_1^{\#} = a^{\#} \overset{n}{\Lambda} \varphi_2^{\#}$, it is easy to see that $\varphi_1^{\#}\varphi_2^{\#} = 0$. By theorem 1, $\varphi_1^{\#}$ reduces to φ_1 modulo (x) , so lemma 2 applies and shows that $\mathrm{Coker}(\varphi_1^{\#})$ is a lifting of $\mathrm{Coker}(\varphi_1)$.

We now return to the problem with which we began: what can one say about the relations between the maps φ_i in (2) ? If the F_i were vector spaces, the exactness of (2) would be equivalent to a condition on the ranks of the maps φ_i .

In the case of, say, a local ring R , what additional condition or conditions must be imposed in order to ensure exactness of a sequence of free R-modules? In order to answer this, we introduce some terminology and notation.

Let R be a commutative ring and $\varphi: F \longrightarrow G$ a map of finitely generated free R-modules. We define the <u>rank of φ</u> to be the largest integer r such that ${}^r\Lambda\varphi: {}^r\Lambda F \longrightarrow {}^r\Lambda G$ is not zero. If $r = \mathrm{rank}\ (\varphi)$, we denote by $I(\varphi)$ the ideal of R generated by the minors of φ of order r .

The definition of $I(\varphi)$ makes sense if one chooses bases of F and G and writes the matrix associated to φ . It can also be shown that the map $\varphi: F \longrightarrow G$ induces, for every k , a map ${}^k\Lambda G^* \otimes {}^k\Lambda F \xrightarrow{\Delta_k} R$ (where $G^* = \mathrm{Hom}_R(G,R)$) and $I(\varphi) = \mathrm{Im}\ (\Delta_r)$.

With this notation, we can state

Theorem 3 [2]. Let R be a noetherian commutative ring, and

$$\mathbb{F}: 0 \longrightarrow F_n \xrightarrow{\varphi_n} F_{n-1} \longrightarrow \cdots \longrightarrow F_1 \xrightarrow{\varphi_1} F_o$$

a complex of free R-modules. Then $\mathbb{F}$ is an exact sequence if and only if

i) $\operatorname{rank}(\varphi_{k+1}) + \operatorname{rank}(\varphi_k) = \operatorname{rank}(F_k)$ for $k = 1,\dots,n$ and

ii) $I(\varphi_k)$ contains an R-sequence of length k for $k = 1,\dots,n$.

We make the convention that condition ii) is automatically satisfied if $I(\varphi_n) = R$. As a result, when R is a field, condition i) becomes the only (and the usual) condition for the exactness of a complex of vector spaces.

Theorem 3 provides one very quick proof of Theorem 1. For suppose

$\mathbb{F}: 0 \longrightarrow R^{n-1} \xrightarrow{(a_{ij})} R^n \xrightarrow{(y_i)} R \longrightarrow R/I \longrightarrow 0$ is exact. Letting $\varphi_2 = (a_{ij})$ and $\varphi_1 = (y_i)$, we know by Theorem 3 that $\operatorname{rank}(\varphi_2) = n-1$ and $I(\varphi_2)$ contains an R-sequence of length two. But $I(\varphi_2) = (\Delta_i)$ where $\{\Delta_i\}$ are the minors of order $n-1$ of the matrix (a_{ij}). Thus, the sequence $0 \longrightarrow R \xrightarrow{(\Delta_i)} R^n \xrightarrow{\varphi_2^*} R^{n-1}$ is exact since the composition is clearly zero and Theorem 3 applies.

Because the sequence $0 \longrightarrow R \xrightarrow{(y_i)} R^n \xrightarrow{\varphi_2^*} R^{n-1}$ is of order two, we get a map of complexes

$$\begin{array}{ccccccc} 0 & \longrightarrow & R & \xrightarrow{(y_i)} & R^n & \xrightarrow{\varphi_2^*} & R^{n-1} \\ & & \downarrow a & & \Big| & & \Big| \\ 0 & \longrightarrow & R & \xrightarrow{(\Delta_i)} & R^n & \xrightarrow{\varphi_2^*} & R^{n-1} \end{array}$$

and thus $y_i = a\Delta_i$ for $i = 1,\dots,n$.

Although Theorem 3 provides us with a nice proof of Theorem 1, it is not clear how that gets us any further into the problem of determining relations among the maps of an arbitrary finite free resolution. In [4], we have exploited Theorem 3 to obtain the following two results:

Theorem 4 [4, Th. 3-1]. Let R be a noetherian ring, let (2) be an exact sequence of free R-modules, and let $r_k = \operatorname{rank}(\varphi_k)$. Then for each k, $1 \leq k \leq n$, there exists an unique homomorphism $a_k : R \longrightarrow \overset{r_k}{\Lambda} F_{k-1} \approx \overset{r_{k-1}}{\Lambda} F^*_{k-1}$ such that

i) $a_n = \overset{r_n}{\Lambda} \varphi_n : R = \overset{r_n}{\Lambda} F_n \longrightarrow \overset{r_n}{\Lambda} F_{n-1}$

ii) for each $k < n$, the diagram

$$\begin{array}{ccc} \overset{r_k}{\Lambda} F_k & \xrightarrow{\overset{r_k}{\Lambda}\varphi_k} & \overset{r_k}{\Lambda} F_{k-1} \\ & {\scriptstyle a^*_{k-1}} \searrow \quad \swarrow {\scriptstyle a_k} & \\ & R & \end{array}$$

commutes.

Using the maps a_k, we may define maps $a'_k : \overset{r_k-1}{\Lambda} F_{k-1} \longrightarrow F^*_{k-1}$ to be the composite:

$$\overset{r_{k-1}-1}{\Lambda} F_{k-1} = R \otimes \overset{r_k+1}{\Lambda} F^*_{k-1} \xrightarrow{a_k \otimes 1} \overset{r_k}{\Lambda} F_{k-1} \otimes \overset{r_k+1}{\Lambda} F^*_{k-1} \xrightarrow{n} F^*_{k-1}$$

where $n : \overset{r_k}{\Lambda} F_{k-1} \otimes \overset{r_k+1}{\Lambda} F^*_{k-1} \longrightarrow F^*_{k-1}$ is the usual action of ΛF_{k-1} on ΛF^*_{k-1} (see [1]).

Theorem 5. [4, Th. 6.1]. Let notation and hypothesis be as in Theorem 4. Then for $k \geq 2$, there exist maps $b_k : F^*_k \longrightarrow \overset{r_{k-1}}{\Lambda} F_{k-1}$ making the diagram

$$\begin{array}{ccc} \overset{r_{k-1}}{\Lambda} F_k & \xrightarrow{\overset{r_k-1}{\Lambda}\varphi_k} & \overset{r_k-1}{\Lambda} F_{k-1} \\ & {\scriptstyle a'_{k-1}} \searrow \quad \swarrow {\scriptstyle b_k} & \\ & F^*_k & \end{array}$$

commute.

Theorems 4 and 5 give us fairly strong information about the relations of the minors of orders r_k and r_k-1 of the maps φ_k in a finite free resolution (2). If we have the exact sequence

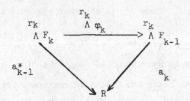

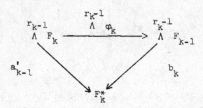

$$F;\ 0 \longrightarrow R^{m-2} \xrightarrow{\varphi_3} R^m \xrightarrow{\varphi_2} R^3 \xrightarrow{\varphi_1} R ,$$

then $r_1 = 2$, $r_1 -1 = 1$, and it is possible to use Theorems 4 and 5 to express the maps φ_1 and φ_2 completely in terms of the minors of φ_3 of order m-2 (see [4]). Consequently just as Theorem 1 was applied to the lifting problem for cyclic modules of homological dimension 2, Theorems 4 and 5 may be applied to show that cyclic modules R/I of homological dimension 3 may be lifted, provided I is generated by 3 elements.

In the general case of a cyclic module R/I of homological dimension 3, our results applied to a resolution of R/I:

$$0 \longrightarrow R^p \xrightarrow{\varphi_3} R^m \xrightarrow{\varphi_2} R^n \xrightarrow{\varphi_1} R \longrightarrow R/I \longrightarrow 0$$

only give us information about the minors of order n-1 and n-2 of φ_1 . One might therefore hope that further information about the lower order minors of φ_1 and φ_2 might be obtained from information about the lower order minors of φ_3 . This idea, of course, is completely demolished when one finds resolutions of the form

$$(5)\quad 0 \longrightarrow R \xrightarrow{\varphi_3} R^n \xrightarrow{\varphi_n} R^n \xrightarrow{\varphi_1} R \longrightarrow R/I \longrightarrow 0,$$

particularly if I contains an R-sequence of length three. If R is a regular local ring, this turns out to be the case precisely when R/I is a Gorenstein ring. It is possible to show, however, that the lifting problem for 5) in the Gorenstein case reduces to a problem of lifting

$$0 \longrightarrow R^p \xrightarrow{\psi_3} R^m \xrightarrow{\psi_2} R^4 \xrightarrow{\psi_1} R \longrightarrow R/J \longrightarrow 0$$

where J is an ideal generated by four elements and contains an R-sequence of length three. Since $p = m-4+1 = m-3$, the map ψ_3 will have lower order minors than those of order p, provided $m - 3 > 1$ i.e. $m > 4$. The question arises: is it possible to have an ideal I in a regular local ring R such that

i) R/I is a Gorenstein ring
ii) $hd_R\ R/I = n$ and
iii) I is minimally generated by n + 1 elements?

In [3], we show that this situation cannot arise. In fact, the question arises as to what restrictions there are on the number of generators of an ideal I in a regular local ring R when R/I is Gorenstein. If dim R = 3 and dim R/I = 0, we have found such ideals I generated by five, seven, and nine elements, but none that are generated by an even number of elements.

This appears to be tied up with questions about the possible (or probable) skew-symmetry of the map φ_2 in the resolution of such an ideal:

$$0 \longrightarrow R \xrightarrow{\varphi_3} R^n \xrightarrow{\varphi_2} R^n \xrightarrow{\varphi_1} R \longrightarrow R/I \longrightarrow 0$$

A computer program worked out by R. Zibman for the Brandeis PDP-10 computer has provided numerous examples of such ideals I and may yield some helpful information on the problem. In any event, it seems evident that our problem in linear algebra has many ramifications, the nature of which we are just beginning to discover.

Bibliography

1. N. Bourbaki, Elements de Mathematiques, Algebre, Ch. III, Hermann, 1970.
2. D.A. Buchsbaum and D. Eisenbud, What makes a complex exact, J. Alg. to appear.
3. ________ ________, Remarks on ideals and resolutions, Symposia Math. to appear.
4. ________ ________, Some structure theorems for finite free resolutions, To appear.
5. L. Burch, On ideals of finite homological dimension in local rings, Proc. Cam. Phil. Soc. 64 (1968) 941-946.
6. D. Hilbert, Uber die Theorie der algebraischen Formen, Math. Ann. (1890), 473-534.
7. I. Kaplansky, Commutative Rings, Allyn and Bacon, 1970.

MAXIMAL IDEALS IN POLYNOMIAL RINGS

Edward D. Davis and Anthony V. Geramita
Dept. of Math., SUNYA, Albany, N. Y. 12222, USA
Dept. of Math., Queen's Univ., Kingston, Ont., Canada

This note is concerned with estimates of the minimal number of generators for maximal ideals in polynomial rings -- estimates extending the well known facts of the case wherein the coefficient ring is a field.

As ever, all rings are commutative with $1 \neq 0$. Let $\nu(I)$ denote the minimal number of generators of the ideal I. (This number may of course be infinite.) We are here interested in some easily obtained, but in many cases sharp estimates of $\nu(M)$ for M a maximal ideal of the polynomial ring $S = R[X_1,\ldots,X_n]$. A well known consequence of Hilbert's Nullstellensatz is that $\nu(M) = n$ if R is a field [8]. We shall obtain a generalization for arbitrary R and as corollary the sharpest possible result for R a regular Noetherian Hilbert ring, for example, a Dedekind domain with infinitely many maximal ideals.

From now on R, S and M are as above, and $P = M \cap R$. We need one fact from Kaplansky's polished rendering of the Krull-Goldman treatment of the Nullstellensatz [4, 6]: R/P is a G-domain [5, Thm. 23]. This means that there exists t in R such that PR_t is maximal. (The reader who cares only for Noetherian rings can manage without this fact.) We need also the upper semicontinuity of ν: For I and Q ideals of a ring A, with I finitely generated and Q prime, there exists t in A-Q such that $\nu(IA_t) = \nu(IA_Q)$.

THEOREM. With R, S, P, M as above, assume $\nu(P)$ finite. Then:

(1) For P maximal, $\nu(M) \leq \nu(P)+n$.

(2) In any case, $\nu(M) \leq \nu(PR_P)+n+1$.

(3) For P maximal and n positive, $\nu(M) \leq \nu(PR_P)+n$.

Proof. For P maximal S/PS is a polynomial ring over the field R/P; (1) is then an immediate consequence of the known case wherein the coefficient ring is a field. As for (2), observe that the upper semicontinuity of ν and the fact that R/P is a G-domain guarantee the existence of t in R such that PR_t is maximal and $\nu(PR_t) = \nu(PR_P)$. By (1) then, $\nu(MS_t) \leq \nu(PR_P)+n$. Take m in M comaximal with t and let I be the ideal of S generated by m and $\nu(PR_P)+n$ elements of M which generate MS_t. Then I = M because $IS_Q = MS_Q$ for every maximal ideal Q of S. As for (3), note that the Nullstellensatz applied to the polynomial ring S/PS shows that $M \cap R[X_1]$ is maximal. This fact and (1) show that it suffices to take n = 1. In this case M/PS is principal; say f generates M mod PS. Taking t as in the proof of (2) we have that $\nu(PA) \leq \nu(PR_P)$ for any R-algebra A such that tA = A. Let g = tf-p, where 1 = tr+p ($r \in R$, $p \in P$). Then since rg = f-pf-pr, M = gS+PS. Hence for A = S/gS, it suffices to check that tA = A: 1 = tr+p = tr+tf*, where f* denotes the canonical image of f in A.

REMARKS. 1. Taking n = 0 in (2) gives: For Q a finitely generated maximal ideal of a ring A, $\nu(Q) \leq \nu(QA_Q)+1$. (Observe that this fact is a consequence of "upper semicontinuity" alone. And somewhat more generally: For I finitely generated and A/I semilocal, $\nu(I) \leq 1+\max\{\nu(IA_Q) \mid Q \text{ maximal}\}$.) 2. One can avoid use of the fact about G-domains in the proof of (2) by assuming $\nu(M)$ finite: Localize at P; apply (1) to conclude that $\nu(MS_M) \leq \nu(PR_P)+n$; then apply Remark 1. 3. In general no better estimate than (2) is possible: Take R = S a Dedekind domain and M a nonprincipal maximal ideal. 4. Remark 3 shows that one cannot in general weaken the hypothesis on n in (3).

Nor can one weaken the hypothesis on P: one easily creates examples for n = 1 wherein R is a 1-dimensional local domain and M is a non-principal maximal ideal for which P = 0.

Assume R Notherian and so that $\dim(S_M) = \dim(R_P)+n$. If S_M is regular (iff R_P is), P maximal and n > 0, then $\nu(M) = \dim(S_M)$; whence M is generated by a regular sequence [2] -- not necessarily so for P nonmaximal (Remark 4). So if we insure P's maximality by taking R a Hilbert ring, there results a theorem well known for R a field:

COROLLARY. Every maximal ideal of a polynomial ring in a positive number of indeterminates over a regular Noetherian Hilbert ring is generated by a regular sequence.

Question: Is this result valid without the "Hilbert ring" proviso?
Answer: Yes in certain semilocal cases: dim(R) = 1 (Endo [1]); n = 1, dim(R) = 2 (Geramita [3]). A special case of the corollary -- R a Dedekind domain with infinitely many maximal ideals -- was discovered independently by Swan, by Gilmer, by Geramita and by W. Heinzer and Davis. The proof of (3) is essentially the Heinzer-Davis treatment.

Suppose now that R is a Dedekind domain and n = 1. For Q any proper ideal of S with $Q \cap R$ nonzero, one routinely adapts the proof of (3) to show that $\nu(Q) \leq 2$ if $Q/(Q \cap R)S$ is principal -- e.g., if $Q \cap R$ is the product of distinct maximal ideals. For I a nonzero proper ideal of S, I = HJ, where H is the intersection of the height 1 primary components of I and J = (I:H) -- J is "the" 0-dimensional component of I. From this decomposition one deduces -- "divide" by H -- that I is S-isomorphic to an ideal Q = JK, where $K = (K \cap R)S$ and is comaximal with J. It follows that $Q/(Q \cap R)S$ is principal if $J/(J \cap R)S$ is. Hence $\nu(I) \leq 2$ if, for example, $J \cap R$ is the product of distinct maximal ideals. A special case of this has been observed by Gilmer: $\nu(I) \leq 2$ for I a radical ideal. For R a PID these remarks are special cases of a theorem of Serre [7]: $\nu(I) \leq 2$ if $\nu(IS_N) \leq 2$ for every maximal N. Question: How much of Serre's theorem survives for arbitrary Dedekind domains? Observe that the validity of the theorem in the more general setting is an immediate corollary of the Eisenbud-Evans conjectured sharpening of the Forster-Swan bound on the number of generators of an S-module [0].

REFERENCES

[0] Eisenbud, D. and Evans, E. G. Generating modules efficiently over polynomial rings, these proceedings.

[1] Endo, S. Projective modules over polynomial rings, J. Math. Soc. Japan 15 (1963) 339-352.

[2] Davis, E. D. Ideals of the principal class, R-sequences and ..., Pac. J. Math. 20 (1967) 197-209.

[3] Geramita, A. V. Maximal ideals in polynomial rings, to appear.

[4] Goldman, O. Hilbert rings and the Hilbert Nullstellensatz, Math. Z. 54 (1951) 136-140.

[5] Kaplansky, I. Commutative Rings, Allyn and Bacon, Boston (1970).

[6] Krull, W. Jacobsonscher Ringe, Hilbertscher Nullstellensatz, Dimensiontheorie, Math. Z. 54 (1951) 354-387.

[7] Serre, J-P. Sur les modules projectifs, Seminaire Dubreil-Pisot No. 2 (1960/61).

[8] Zariski, O. Foundation of a general theory of birational correspondences, Trans. A.M.S. 53 (1943) 490-542.

A CANCELLATION PROBLEM FOR RINGS

PAUL EAKIN and WILLIAM HEINZER

0. Introduction. Suppose A and B are rings. Let us say that A and B are stably equivalent if there is an integer n such that the polynomial rings $A[X_1,...,X_n]$ and $B[Y_1,...,Y_n]$ are isomorphic. A number of recent investigations [CE], [AEH], [EK], [BR], [H] are concerned with the study of this equivalence. The most obvious question one might ask is:

(0.1) Cancellation Problem. Suppose A is a ring. What conditions on A guarantee that for any ring B, A is stably equivalent to B only if A is isomorphic to B?

This cancellation problem was first investigated in [CE] and later in [AEH], [EK], [BR] and [H]. In [H], Hochster provides an example of two stably equivalent rings which are not isomorphic. In addition to providing an example, [H] also draws our attention to the connection between the cancellation problem for rings and the well-studied cancellation problem for modules.

In the study of the cancellation problem for rings the following concepts have proved useful.

(0.2) A ring A is said to be **invariant** provided that whenever A is stably equivalent to some ring B, then A is isomorphic to B.

(0.3) A ring A is said to be **strongly invariant** provided that whenever $\sigma: A[X_1,...,X_n] \to B[Y_1,...,Y_n]$ is an isomorphism of polynomial rings, then $\sigma(A) = B$.

Obviously, strongly invariant rings are invariant. The most obvious example of a strongly invariant ring is the integers Z. On the other hand there are invariant rings which are not strongly invariant. The following is a result from [AEH].

(0.4) Suppose A is a domain of transcendence degree one over a field, then A is invariant. Moreover A is strongly invariant unless there is a field k such that $A = k[X]$.

This result tells us that if k is a field then $A = k[X]$ is an invariant ring. However A is not strongly invariant, for if $A = k[X]$ and $B = k[Y]$ then the identity is an isomorphism of $A[Y]$ onto $B[X]$ which does not take A onto B.

A fair amount can be said about rings of Krull dimension one which are not strongly invariant [AEH], but this case is not completely settled. In particular there is a gap in our knowledge of the case when A is a Dedekind domain. In section five we have a discussion of the cancellation problem for Dedekind domains and present some questions whose affirmative answer would affirmatively settle the cancellation problem for all one dimensional noetherian domains.

One way to approach the cancellation problem for a particular ring A is to write

$$A[X_1,...,X_n] = B[Y_1,...,Y_n]$$

for some ring B and look at $D = A \cap B$. In the case of domains, if the transcendence degree of A over D is at most one and S denotes the non zero elements of D, then either $A = B$ or $A_s \cong B_s \cong k[X]$ where k is the quotient field of D. This follows from the result on domains of transcendence one over a field mentioned above, with information like $A_s = k[t]$, a bit of additional hypothesis will often yield results like $A \approx B \approx D[X]$. When the transcendence of A over D is greater than one, the problem becomes considerably more complex. In particular, the cancellation problem for $\mathbb{C}[X,Y]$, the polynomials in two variables over the complex numbers is not settled. C.P. Ramanujam has taken a particularly sophisticated topological approach to this question [R]. He characterizes the complex 2-space as a non singular, contractible algebraic surface which is simply connected at infinity. To our knowledge, this has not yielded a solution to the cancellation problem for $\mathbb{C}[X,Y]$, $\mathbb{C}$ the complex numbers.

We begin this article by examining Hochster's example from [H]. We observe that this provides a nice introduction to an interesting class of rings, called **locally polynomial rings** in [ES]. (Given a ground ring R and an algebra A over R, we say that A is locally a polynomial ring over R if $A_p = A \otimes_R R_p$ is a polynomial ring over R_p for every prime p of R). Over a fairly substantial collection of rings we give a number of characterizations of locally polynomial rings "in one variable." In particular we show that if D is a domain, the affine (ie. finitely generated ring extension) locally polynomial rings in one variable over D are precisely the symmetric algebras of the invertible ideals of D. We then use this result to indicate an approach to the cancellation problem for $k[X,Y]$, k an algebraically closed field.

Before proceeding let us make the usual stipulations: all rings are commutative with identity and all modules are unital.

1. Symmetric Algebras and Non Invariant Rings. In this section we give Hochster's example of a non invariant ring. This example, in addition to showing that stable equivalence of rings is not a trivial relationship (i.e. isomorphism) also serves to remind "commutative algebraists" that its nice to know some topology-especially the theory of vector bundles.

Let us recall the definition and basic properties of the symmetric algebra of a module.

(1.1) Definition. Let R be a ring and M an R module. A symmetric algebra of M over R is a pair (φ, S(M)) consisting of an R isomorphism φ and an R algebra S(M) such that φ: M → S(M) and such that if σ is any R-homomorphism of M into an R algebra A, then there is a unique R-homomorphism h such that the following diagram commutes

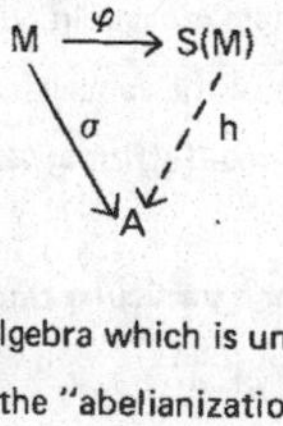

Every module has a symmetric algebra which is unique up to an R-isomorphism [C]. In fact, if M is an R module, S(M) is R isomorphic to the "abelianization" of the tensor algebra of M over R [C]. Another realization of the symmetric algebra of a module which is often quite computable is the following.

Let M be a module and write

$$0 \to K \to F \overset{\eta}{\to} M \to 0 \qquad \text{exact}$$

where F is a free module, say $F = \sum_{i \in I} RZ_i$. Let $\{X_i\}_{i \in I}$ be indeterminates over R and consider $J(K) = \{\Sigma\, r_i X_i \in R[\{X_i\}] \mid \Sigma\, r_i Z_i \in K\}$. Let $\mathfrak{a}$ denote the ideal in $R[\{X_i\}_{i \in I}]$ generated by J. Then $R[M] = R[\{X_i\}_{i \in I}]/\mathfrak{a}$ is the symmetric algebra of M over R. To see this one observes that R[M] is a graded, homogeneous ring over R whose module of 1-forms is isomorphic to M. Thus we have an R-homomorphism h such that

$$\begin{array}{ccc} M & \overset{i}{\to} & R[M] \\ \downarrow & \overset{\varphi}{} \nearrow h & \\ S(M) & & \end{array} \qquad \text{commutes.}$$

Let $z_i = \eta(Z_i)$. Then we have a mapping

$$R[\{X_i\}_{i \in I}] \overset{\ell}{\to} S(M) \qquad \text{given by} \qquad x_i \to \varphi(z_i).$$

Since $J \subset \operatorname{Ker} \ell$ we have a well defined mapping $R[M] \overset{\bar{\ell}}{\to} S(M)$ which is surjective. Since $h \circ \bar{\ell} = id$ we have R[M] is R-isomorphic to S(M).

Using the above, or referring to [C] one can easily see the following well known facts.

(1.2) Lemma. Let R be a ring, then

(i) If F is a free R module on a set of cardinality I, S(F) is R isomorphic to the polynomials over R in a set of indeterminates having cardinality I.

(ii) If A and B are R modules then $S(A \oplus B)$ is R isomorphic to $S(A) \otimes_R S(B)$.

Recall that two R-modules M and N are **stably equivalent** if there is a free module F such that $M \oplus F \approx N \oplus F$. In this case, if F is free on n generators then, taking symmetric algebras we have

$S(M) \otimes_R S(F) \cong S(M \oplus F) \cong S(N \oplus F) \cong S(N) \otimes_R S(F)$. But

$$S(F) \approx R[X_1, \ldots, X_n]. \qquad \text{Hence we have}$$

$$S(M)[X_1, \ldots, X_n] \approx S(M) \otimes_R R[X_1, \ldots, X_n] \approx S(N) \otimes_R R[X_1, \ldots, X_n] \approx S(N)[X_1, \ldots, X_n].$$

Thus stably equivalent modules give rise to stably equivalent rings. This provides the basic idea of Hochster's example: find two stably equivalent modules which are not isomorphic, then prove that their respective symmetric algebras are not isomorphic.

(1.3) Lemma. Let R be a ring and M and N two finitely generated R modules. If S(M) and S(N) denote the respective symmetric algebras then S(M) is R isomorphic to S(N) if and only if M is isomorphic to N.

Proof. Clearly $M \cong N :\Rightarrow:$ S(M) is R-isomorphic to S(N). Conversely, suppose S(M) is R-isomorphic to S(N), say by an R-isomorphism φ. Now S(M) is a homogenous graded ring whose module of 1-forms is isomorphic to M. Identify M with these 1-forms and let $\{\eta_i\}_{i\in I}$ be a basis for M. Then write $\varphi(\eta_i) = \sum_{j=0} \alpha_{ij} \in S(N)$. Here α_{ij} is a homogenous element of S(N) of degree j. Since φ is an R-isomorphism, we have

$$\varphi(\eta_i - \alpha_{oi}) = \sum_{j=1} \alpha_{ij}\,.$$

Denote this element by γ_i. Then since $S(M) = R[\{\eta_i\}_{i\in I}] = R[\{\eta_i - \alpha_{oi}\}]$ we have $S(N) = R[\{\gamma_i\}_{i\in I}]$. By grading it follows that $\{\alpha_{1i}\}_{i\in I}$ generates the 1-forms of S(N), which is a module isomorphic to N. There is then a well defined surjective R homomorphism

$$\sigma: M \to N \qquad \text{given by} \qquad \sigma(\eta_i) = \alpha_{i1}$$

which takes M onto N. By symmetry there is a surjection $\sigma': N \to M$. But $\sigma' \circ \sigma$ is a surjective mapping of M onto itself. Thus $\sigma' \circ \sigma$ is an isomorphism [V] and so is σ.

Thus, by this lemma we see that if M and N are finitely generated, stably equivalent, non isomorphic modules, then S(M) and S(N) cannot be R-isomorphic. The idea now is to find two such modules and prove that any supposed isomorphism of the symmetric algebras would (essentially) have to be an R isomorphism. To this end we recall the following from [S].

Let k denote the real numbers and $R = k[X_o, \ldots, X_n] / \sum_{i=0}^{n} X_i^2 - 1 = k[x_0, x_1, \ldots, x_n]$. Let F denote the free module on generators $e_o, \ldots, e_n$ and define $\sigma: F \to F$ by

$$\sigma(e_i) = x_i(\Sigma\, x_i e_i)\,.$$

Then one has

$$0 \to P \to F \overset{\sigma}{\to} (\Sigma\, x_i e_i)R \to 0 \quad \text{exact.}$$

Hence $P \oplus R \approx R^{n+1} = R^n \oplus R$. Moreover $P \approx R^n$ (ie P is free) if and only if n=1,3,7.

The notation of this statement holds for the next two lemmas.

(1.4) Lemma. The ring R above is strongly invariant.

Proof. Suppose $\gamma': R[X_1, \ldots, X_n] \to B[Y_1, \ldots, Y_n]$. We may assume $R[X_1, \ldots, X_n] = B[Y_1, \ldots, Y_n]$ and show R = B. Suppose

$$x_i = b_{oi} + b_{1i} Y_n + \cdots + b_{ki} Y_n^k \quad \text{where } b_{ij} \in B[Y_1, \ldots, Y_{n-1}].$$

Then $\Sigma x_i^2 = 1$ implies $\Sigma b_{k,i}^2 = 0$ if k is greater than zero. But $R[X_1, \ldots, X_n]$ is formally real (ie. 0 cannot be written non trivially as a sum of squares there. Thus $x_i = b_{oi}$ for each i and $x_i \in B[Y_1, \ldots, Y_{n-1}]$. By further reduction, $x_i \in B$ for each i and hence $R \subset B$. But it now follows easily that R = B.

(1.5) Lemma. If φ: S(P) $\to R[X_o, \ldots, X_{n-1}]$ then up to an automorphism of R, φ is an R isomorphism.

Proof. By adding an indeterminate and extending trivially we have

$$\varphi: S(P)[Y] \to R[X_o,\ldots,X_{n-1}][Z].$$

Since $S(P)[Y] = R[Z_o,\ldots,Z_n]$, one concludes that, up to an automorphism σ of R, φ is an R isomorphism. However one then has

$$S(P) \overset{\varphi}{\to} R[X_o,\ldots,X_{n-1}] \overset{\sigma^{-1}}{\to} R[X_o,\ldots,X_{n-1}]$$

and $\sigma^{-1}\circ\varphi$ is an R isomorphism. □

Thus, with the notation of the previous lemmas, if $R[X_1,\ldots,X_n]$ is isomorphic to S(P), then $P \cong R^n$ by (1.3). So for $n \neq 1,3,7$ we see that the rings S(P) and $R[X_1,\ldots,X_n]$ are non isomorphic, stably equivalent rings.

(1.6) Remark. **In case n = 2 in the above, the integral domain R has the following properties**

(i) R is strongly invariant

(ii) R[X] is invariant, but not strongly invariant

(iii) R[X,Y] is not invariant.

Proof. We have already seen (i) and (iii). To see (ii) we observe that the proof of (1.4) actually shows somewhat more. We have seen that if $R[X_1,\ldots,X_n] = B[Y]$. Then $R \subset B$. We can thus appeal to the following result from [AEH].

Let D be a unique factorization domain (UFD) and let $\{X_i\}$ be indeterminates over A. If R is a UFD such that $D \subset R \subset D[\{X_i\}]$ and such that R has transcendence degree one over D, then R is a polynomial ring over D.

Remark (1.6) has to be considered when one gets overly optimistic about the prospects of proving k[X,Y] invariant, k a field.

2. Locally Polynomial Rings. Let D be a ring and A an algebra over D. If for every prime $p \subset D$, $A_p = D_p \otimes_D A$ is a polynomial ring over D, we say that A is **locally a polynomial ring** over D. These rings are investigated in [ES]. Since the formation of symmetric algebras respects change of ring, we find that the symmetric algebras of locally free modules provide plenty of examples of locally polynomial rings.

(2.1) Lemma. **If M_1 and M_2 are modules over a ring R such that**

$$S(M_1)[X_1,\ldots,X_n] \underset{R}{\cong} S(M_2)[Y_1,\ldots,Y_n]$$

(ie. the algebras are R-isomorphic). Then for every prime $p \subset R$, $S(M_1) \otimes R_p \underset{R_p}{\cong} S(M_2) \otimes R_p$.

Proof. By lemma (1.3) we have that M_1 and M_2 are stably equivalent modules. Thus $M_1 \otimes_R R_p$ and $M_2 \otimes_R R_p$ are stably equivalent R_p modules. Thus $M_1 \otimes_R R_p$ is R_p isomorphic to $M_2 \otimes_R R_p$ by a theorem of Bass which assures us that stably equivalent modules over a local ring are isomorphic [B]. Thus $S(M_1 \otimes_R R_p) = S(M_1) \otimes_R R_p$ is R_p isomorphic to $S(M_2) \otimes_R R_p = S(M_2 \otimes_R R_p)$.

With reference to the preceeding lemma we see that if $S(M_1)$ is a domain and $S(M_1)$ is stably equivalent to $S(M_2)$, then even though $S(M_1)$ may not be isomorphic to $S(M_2)$, they still have isomorphic quotient fields. Referring to Hochster's example then, we see that this is not a counterexample to the following "bi-rational cancellation problem".

Suppose A and B are domains and $A[X] \cong B[Y]$ is an isomorphism of polynomial rings. Are the quotient fields of A and B isomorphic?

This question is considered in [AEH]. It is certainly related to the following, well studied question of Zariski:

Zariski Problem. Let K and K' be finitely generated fields over a field k. Assume that simple transcendental extensions of K and K' are k-isomorphic to each other. Does it follow that K and K' are k-isomorphic to each other?[1]

3. Some Interesting Types of Ring Extensions. In studying the cancellation problem we have been led to consider the following possible relationships between a ring R and its extension A.

(1) A is the symmetric algebra of a projective R module

(2) A is projective in the category of R algebras. That is, given a diagram of R-algebras and R homomorphisms.

$$\begin{array}{ccccc} & & A & & \\ & & \downarrow & & \\ B & \rightarrow & C & \rightarrow & 0 \end{array}$$

with the bottom row exact, there is a R homomorphism $\theta: A \rightarrow B$ which makes the diagram

$$\begin{array}{ccccc} & \theta \swarrow & A & & \\ & & \downarrow & & \\ B & \rightarrow & C & \rightarrow & 0 \end{array}$$

commute.

(3) A is a retract of a polynomial ring over R. That is, there is an idempotent endomorphism of some polynomial ring $R[\{X_i\}_{i\epsilon I}]$ which has A as its range.

(4) There exists an R-algebra A^* such that $A \otimes_R A^*$ is a polynomial ring over R.

(5) There are indeterminates $\{X_\alpha\}$ and $\{Y_\beta\}$ such that $R[\{X_\alpha\}] = A[\{Y_\beta\}]$.

(6) There exist indeterminates $\{X_\alpha\}$ such that $R \subset A \subset D[\{X_\alpha\}]$ and A is an inert subring of $R[\{X_\alpha\}]$. That is, if a,b are non zero elements of $R[\{X_\alpha\}]$ and ab ϵ A then a ϵ A and b ϵ A.

---

1 This question is studied by Nagata in [TVR]. He notes there that it was raised by Zariski at a 1949 Paris Colloquium on algebra and the theory of numbers.

(7) A is locally a polynomial ring over R.

At present the relationships among these properties is not clear. However, some connections do follow readily.

(3.1) Lemma. With regard to the above conditions we have the following implications:

$$\begin{array}{ccccccc} (7) & \Leftarrow & (1) & \Rightarrow & (2) & \Leftrightarrow & (3) \\ & & \Downarrow & & & \nearrow & \\ & & 4 & \Rightarrow & 5 & & \end{array}$$

In case R is a domain we have (5) ⇒ (6), and if A is an affine ring of transcendence degree one over the domain R then (7) ⇒ (1).

Proof. (1) ⇒ (7) Let P be a projective R module and Q a prime of R. Then $S_R(P) \otimes_R R_Q = S_{R_Q}[P \otimes_R R_Q]$. But $P \otimes_R R_Q$ is free, hence $S_{R_Q}[P \otimes_R R_Q]$ is a polynomial ring.

(2) ⇔ (3) The proof is the same as showing that projective modules are retractions of free modules.

(1) ⇒ (4) Let A = S(P) where P is a projective R module and let P* be an R module such that $P \oplus P^*$ is free. Then $S(P) \otimes_R S(P^*) = S(P \oplus P^*)$ is a polynomial ring over R.

(4) ⇒ (5) Suppose $A \otimes_R A'$ is a polynomial ring over R. Then look at

$$A \otimes A' \otimes A \otimes A' \otimes A \otimes \cdots\cdots$$

associated as

$$(A \otimes A') \otimes (A \otimes A') \otimes (A \otimes A') \otimes \cdots\cdots = S$$

and as

$$A \otimes (A' \otimes A) \otimes (A' \otimes A) \otimes (A' \otimes A) \otimes \cdots = A \otimes S.$$

We observe that S is the polynomials in some (large) numbers of variables over R. Hence $A \otimes S$ is a ring of polynomials over A which is what was to be shown

(5) ⇒ (3) Obvious since A is a retract of $A[\{X_\alpha\}]$.

In the case of a domain (5) ⇒ (6) is obvious since then (and only then), A is an inert subring of $A[\{X_\alpha\}]$.

We will be done if we show (7) ⇒ (1) under the hypothesis that A is a domain of transcendence degree one and an affine ring over R. For convenience, this is decomposed into a few lemmas.

(3.2) Lemma. Let R be a domain and A an affine ring over R which is locally a polynomial ring in one variable over R. Then there exists a set $\{x_1, \ldots, x_n\} \subset A$ such that $A = R[x_1, \ldots, x_n]$ and given any prime $P \subset R$, $A_P = R_P[x_i]$ for some i, $1 \leqslant i \leqslant n$ (i may depend upon the choice of P).

Proof. Let $\{P_\alpha\}_{\alpha \in I}$ denote the primes of R and x_α an element of A such that $A_{P_\alpha} = R_{P_\alpha}[x_\alpha]$. Then one has $R[\{x_\alpha\}_{\alpha \in I}] = R$. This follows since we can show that, as R modules they are locally equal, and hence equal. Since A is affine over R we may assume $A = R[x_1,\ldots,x_n]$ where $x_i \in \{x_\alpha\}_{\alpha\in I}$. Moreover, for each prime $P \subset R$, $A_p = R_p[x_i]$ for some $x_i \in \{x_1,\ldots,x_n\}$. For let x_p be a local generator at the prime P. Then for each i

$$x_i = \alpha_i\, x_p + \beta_i \qquad \alpha_i, \beta_i \in K = \text{quotient field of } R\,.$$

By uniqueness of representation, $\alpha_i, \beta_i \in R_p$. Residually mod P we have

$$A_p/pA_p = R_p/pR_p\,[\bar{x}_1,\ldots,\bar{x}_n] = k[\bar{x}_p] = k[\bar{x}_1,\ldots,\bar{x}_n]$$

where k is the field R_p/pR_p. Since $\bar{x}_i = \bar{\alpha}_i\,\bar{x}_p + \bar{\beta}_i$ one of the $\bar{\alpha}_i$ is not zero, (ie one of the α_i is a unit in R_p, say α_1. Then one has $R_p[x_p] = R_p[x_1]$. □

(3.3) Lemma. Let R be a domain and $A = R[x_1,\ldots,x_n]$ a transcendental extension of R such that for any prime $P \subset R$, $A_p = R_p[x_i]$ for some i, $1 \leq i \leq n$. Let K denote the quotient field of R and fix some Y such that $R_{(o)} = K[Y]$. Then one can write

$$x_i = a_i\, Y + b_i \qquad \text{with } a_i, b_i \in K.$$

The R module $\sum_{i=1}^{n} R\, a_i$ is an invertible fractionary ideal of R.

Proof. Let $\mathfrak{m}$ be a maximal ideal of R. Then $R_{\mathfrak{m}}[X_1,\ldots,X_n] = R_{\mathfrak{m}}[X_i]$ for some i. Hence

$$x_1 = \frac{a_1}{a_i}\, x_i + \beta_1$$

$$\vdots$$

$$x_n = \frac{a_n}{a_i}\, x_i + \beta_n\,.$$

By uniqueness of representation $\frac{a_j}{a_i}, \beta_j \in R_{\mathfrak{m}}$. Thus $(a_1,\ldots,a_n)\, R_{\mathfrak{m}} = a_i\, R_{\mathfrak{m}}$. So $(a_1,\ldots,a_n)$ is locally principal and thus invertible.

(3.4) Lemma. With the notation of (3.3) let $\mathfrak{a}$ denote the invertible ideal $(a_1,\ldots,a_n)$. Then A is R isomorphic to $S(\mathfrak{a})$, the symmetric algebra of the projective R-module $\mathfrak{a}$.

Proof. Consider M, the R submodule of A generated by $\{x_1,\ldots,x_n\}$. Represent M as $\langle(a_1Y+b_1),\ldots,(a_nY+b_n)\rangle$. Then there is an obvious natural mapping

$$M \overset{\sigma}{\to} \mathfrak{a} \to 0 \qquad \text{given by}$$

$$x_i \overset{\sigma}{\to} a_i.$$

Since $\mathfrak{a}$ is a projective R module, there is a mapping $\sigma'\colon \mathfrak{a} \to M$ such that $\sigma\sigma' = \mathrm{id}_{\mathfrak{a}}$. Then σ' induces a unique R mapping h such that

$$\begin{array}{ccccc} & & 0 & & \\ & & \downarrow & & \\ 0 & \rightarrow & \mathfrak{a} & \overset{\sigma'}{\rightarrow} & A \\ & & \downarrow & \nearrow_{h} & \\ & & S(\mathfrak{a}) & & \end{array}$$

commutes.

In fact, h is an isomorphism. For let p be any prime of R and let $a_pY + b_p$ be a local generator for R at P. Then let

$$Z_p = \sigma'(a_p) = \sum_{i=1}^{n} \lambda_i\, x_i = \sum_{i=1}^{n} \lambda_i\, (a_iY + b_i) = (\Sigma\, \lambda_i a_i)Y + (\sum_{i=1}^{n} \lambda_i b_i).$$

It turns out that Z_p is a local generator for A at p. For applying σ we have

$$a_p = \sigma(\sigma'(a_p)) = \sigma(\Sigma \lambda_i\, x_i) = \Sigma \lambda_i a_i\, .$$

Thus $Z_p = a_pY + \beta$. Since $Z_p \in A_p = R_p[a_pY{+}b_p]$, $Z_p = \alpha(a_pY{+}b_p) + \delta$ where $\alpha, \delta \in R_p$. Thus $\alpha = 1$ and $Z_p - \delta = a_pY{+}b_p$ so $A_p = R_p[a_pY{+}b_p] \subseteq R_p[Z_p] \subseteq A_p$, and we have equality.

Thus the mapping:

$$(S(\mathfrak{a}))_p \overset{h_p}{\rightarrow} A_p \quad \text{is surjective.}$$

But $S(\mathfrak{a})_p = R_p[T]$ and any R_p surjection of $R_p[T]$ onto $R_p[Z_p]$ would necessarily be an isomorphism. Thus the mapping h is locally an isomorphism therefore its an isomorphism. This completes the proof of the lemma, the theorem now follows immediately. □

(3.5) Theorem. Let A be an affine ring of transcendence degree one over the domain D. Suppose D is locally a UFD, then the conditions 1–7 above are equivalent.

Proof. (6) ⇒ (1) follows in view of the result from [AEH] quoted in (1.6), and since a retract of a UFD is a UFD[1] (2) ⇒ (7) follows from the same result. □

---

1 Ed Enochs called this charming little result to our attention. His argument goes as follows: Let $R \subset S$ be rings, R a retract of the UFD S under the mapping σ. It is easily seen that each element in R is a product of irreducibles, thus one need only show that irreducibles in R are prime. Let $\pi \in R$ be irreducible then factor π in S, $\pi = p_1 \cdots p_k$, each p_i prime. Apply σ to both sides and use the fact that π is irreducible to conclude that for a suitable numbering $\sigma(p_i)$ is a unit, $i \geqslant 2$. Thus $\pi = \mu\sigma(p_1)$, μ a unit in R. Now if $\pi | ab$ in R. Then one can assume $p_1q = a$. Since μ is also a unit in S, $(p_1\mu)(\mu^{-1}q) = a$. Then applying σ we see $\pi(\mu^{-1}\sigma(q)) = a$. Thus $a \in \pi R$ and π is prime.

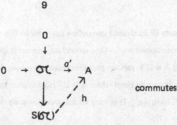

(3.6) Remark. Theorem (3.5) should generalize (at least in the one variable case) to a much larger class of domains than those mentioned here. One would expect most of it to go over to domains D such that D[X] is D invariant. A ring A is D-invariant provided that whenever $A[X_1,\ldots,X_n] \cong B[Y_1,\ldots,Y_n]$ is a D–isomorphism of polynomial rings, then A is D–isomorphic to B. In [AEH] it is shown that for D a locally HCF[1] ring D[X] is D, invariant. It is easy to see that for such rings, all of the implications of (3.5) hold except possibly (2) ⇒ (7).

(3.7) Corollary. Let D be a domain which is locally a UFD and let $\mathfrak{a}$ be an invertible ideal of D. Then $S(\mathfrak{a})$, the symmetric algebra of $\mathfrak{a}$ is D-invariant.[2]

Proof. Suppose $S(\mathfrak{a})[X_1,\ldots,X_n] = B[Y_1,\ldots,Y_n]$ with $D \subset S(\mathfrak{a}) \cap B$. Then localizing everything at a prime p of D, $S(\mathfrak{a})_p = D_p[X_p]$, thus $B_p = D_p[Z_p]$, by the result from [AEH] quoted in the proof of (1.6). Thus B is locally a polynomial ring over D, and since $S(\mathfrak{a})$ is affine over D, so is B. Thus by Theorem (3.5) $B = S(G)$ where G is an invertible ideal of D. If F_n denotes the free D module of rank n, then our hypotheses implies $S(\mathfrak{a} \oplus F_n)$ is D isomorphic to $S(G \oplus F_n)$. Thus $\mathfrak{a} \oplus F_n \approx G \oplus F_n$ and $\mathfrak{a} \approx G$ [K, p.328] and hence $S(\mathfrak{a}) \approx S(G)) = B$.

4. An Approach to Cancellation for $\mathbb{C}[X,Y]$. In this section, we return to what must be regarded as the next fundamental problem in our study; whether $A[T_1,\ldots,T_n] = \mathbb{C}[X_1,\ldots,X_{n+2}]$ implies $A \approx \mathbb{C}[X,Y]$. We offer the following observation.

(4.1) Theorem. Let k be an algebraically closed field and $A[T] = k[X,Y,Z]$ (T and Z represent the same finite number of indeterminates). If k is properly contained in $A \cap k[X,Y]$ then either $A = k[X,Y]$ or there is an f such that $A \cap k[X,Y] = k[f]$. In the latter case, there is a g such that $A = k[f,g]$.

Proof. Let $R = A \cap k[X,Y]$. The two cases obviously depend upon the transcendence degree of R over k. Since R is algebraically closed in A, if the transcendence is two we clearly have equality. If R has transcendence degree one over k then it is known [AEH] that R has the form k[f]. Now k[f] is an inert subring of A, hence any prime element of k[f] already generates a prime ideal in A. Let S denote the non zero elements of k[f], then we have $A_S = k[X,Y]_S[Z]$. Since $A \neq k[X,Y]$, $A_S \neq k[X,Y]_S$. Moreover, k(f) is algebraically closed in A_S and A_S is of transcendence degree one over k(f). Thus by (0.4), $A_S = k(f)[t]$ for some $t \in A$. Now let $V_{(f)}$ denote the rank one discrete valuation ring $k[f]_{(f)}$. Then we have the following situation:

---

1. HCF stands for highest common factor. These are the rings whose groups of divisibility are lattice ordered. Among the rings which locally have this property are the Prufer domains. A student of Hochster has considerably extended the list of domains D such that D[X] is D-invariant.
2. Using (4.9) in [AEH] one can see that HCF can be substituted for UFD.

$V_{(f)} \subset V_{(f)}[t] \subseteq A_{S'} \subset k(f)[t]$, where

$S' = k[f] - f\,k[f]$. We claim that t can be chosen so that $A_{S'} = V_{(f)}[t]$. First we show that we can choose t so that in A/f, $\bar{t}$ is transcendental over k. Let

$$t = b_o + b_1Z + \cdots + b_nZ^n \quad \text{with} \quad b_i \in k[X,Y].$$

Remember that k is algebraically closed and observe that the element t fails to be residually transcendental over k in A/f only if f divides $b_1, \ldots, b_n$ and b_o is residually equal to some $\alpha_o \in k$ (if there is more than one Z you've got to use forms of degree n in place of Z^n). But then f divides $t-\alpha_o$ in k[X,Y,Z], say $fh = t-\alpha_o$. Since $t-\alpha_o \in A$, $h \in A$ and $h = \frac{t-\alpha_o}{f}$ will serve equally well for t. We can repeat this only finitely many times before we come to the point where one of the b_i, $i \geq 1$ is not divisible by f and hence this choice of t is residually transcendental over k mod f. But now we have $A_{S'} = V_f[t]$. For let $a \in A_{S'}$. Then

$$a = \gamma_o + \gamma_1 t + \cdots + \gamma_k t^k \quad \text{where} \quad \gamma_i \in k(f).$$

But each $\gamma_i \in V_{(f)}$, for if $-\ell$ were the least value of any γ_i in the f-adic valuation then

$$f^\ell a = (f^\ell \gamma_o) + (f^\ell \gamma_1)t + \cdots + (f^\ell \gamma_k)t^k. \qquad \text{Then}$$

each $f^\ell \gamma_i$ is an element of $k[f]_{(f)}$ and at least one of them is a unit there. Taking residues mod f, the left side vanishes, and hence the right side becomes an equation of algebraic dependence for t over k mod f, a contradiction. We have thus shown that if p is any prime of R = k[f], then $A_p = A \otimes_R R_p$ is a polynomial ring in one variable over R_p. Thus by Theorem (3.2), A is k[f] isomorphic to the symmetric algebra of an invertible ideal of k[f]. Since k[f] is a PID, this ideal is principal, and the symmetric algebra is k[f] [U]. The image of U under the now extant k[f]-isomorphism provides a g such that A = k[f,g]. □

An interesting corollary is the following:

(4.2) Corollary. Suppose k is an algebraically closed field and k[X,Y,Z] = k[X',Y',Z']. The possibilities for R = k[X,Y] ∩ k[X',Y'] are the following

(i) R = k

(ii) R = k[X,Y]

(iii) R = k[f]. In this case there exist a g and an h such that k[X,Y] = k[f,g] and k[X',Y'] = k[f,h].

Thus the only non-trivial possibility for R is essentially when X = X' and the intersection is k[X].

With regard to the cancellation problem for $\mathscr{C}$[X,Y] we now see the following.

(4.3) Corollary. Let $\mathscr{C}$ denote the complex numbers and suppose $\mathscr{C}$[X,Y,Z] = A[T]. Then A ≈ $\mathscr{C}$[X,Y] if and only if then is an automorphism σ of $\mathscr{C}$[X,Y,Z], say σ(X) = X', σ(Y) = Y', σ(Z) = Z' such that $\mathscr{C}$[X',Y'] ∩ A ≠ k.

The steps in the proof of (4.2) are

(1) Observe that $A \cap k[X,Y] = k[f]$ is an inert subring of A (this is equivalent to the assertion that $f-\lambda$ is prime in A for each $\lambda \in k$)

(2) Argue that $A[k(f)] = k(f)[t]$ for some $t \in A$.

(3) Show that A is locally a polynomial ring in one variable over $k[f]$.

(4) Appeal to (3.5).

In step (3) we use the fact that all operations take place in $k[X,Y,Z]$ to argue that, for each λ, only a finite number of modifications of t are necessary to arrive at a local generator at $k[f]_{(f-\lambda)}$. If we are willing to use a bit more machinery, then we can follow steps (1) and (2) by a global version of step (3), thus eliminating the need for (3.5). In fact, the following is true:

(4.4) Theorem. Let k be an algebraically closed field and A a two dimensional affine ring over k which is a UFD. Suppose there exists an $f \in A$ such that

(i) $k[f]$ is an inert subring of A.

Equivalently; $(f-\lambda)A$ is a prime ideal for every $\lambda \in k$.

(ii) $A[k(f)] = k(f)[t]$

Then there exists $g \in A$ such that $A = k[f,g]$.

Before giving a proof of (4.4) we need to recall some facts about quadratic transformations.

Let $R_{\mathfrak{m}}$ be a local ring and V a valuation ring which dominates R. Let $x_1, \ldots, x_k$ be a minimal generating set for $\mathfrak{m}$ and assume x_k has minimal value in the valuation associated with V. Let $\mathfrak{m}_1$ denote the center of V on $R_1 = R[\frac{x_1}{x_k}, \ldots, \frac{x_{k-1}}{x_k}]$ and consider the local ring $R_{1\mathfrak{m}_1}$. This is a quadratic transform (also called a quadratic dilitation) of R along V. Obviously such a procedure can be iterated. We then speak of a sequence of quadratic transformations of R along V. Let $R < R_1 < R_2 < \cdots < R_n <$ be a sequence of quadratic transformations of R along V. If $\cup R_i = V$ we say that the sequence "reaches" V. If $R_n = V$ for some n, we say that the sequence reaches V in a finite number of steps. In some situations, it is possible to assert that every sequence of quadratic transforms of R along V will reach V in finitely many steps. We need the following special case of a general theorem of Abhyankar [A, p.336].

(4.5) Lemma. Let X and Y be indeterminates over the field k and V a rank one discrete valuation ring such that $k[X,Y] \subset V \subset k(X,Y)$. Suppose that the residue field of V is transcendental over k.[1] Then if p denotes the center of V on $k[X,Y]$, any sequence of quadratic transformations of $k[X,Y]_p$ along V must reach V after finitely many steps.

We can now give a fairly straightforward proof of (4.4).

---

1. In the language of Zariski–Samuel, Vol. II, p.95, V is a prime divisor.

Proof of (4.4). The idea is to choose t as a first approximation to g and argue that only finitely many modifications are necessary to arrive at a "perfect fit." We start out then with $g_1 = t$ and $k[f,g_1] \subset A \subset k(f)[g_1]$. Since A is affine over k, it is easily seen that there is some $\varphi \in k[f]$ such that $k[f,g_1] \subset A \subset k[f,g_1, 1/\varphi]$. Let $h_1,\ldots,h_s$ denote the prime factors of φ. With the possible exception of the $h_1,\ldots,h_s$–adic valuations of A, every essential valuation of A is an essential valuation of $k[f,g_1]$. We claim that we can modify g_1 to get g_2 so that the $h_1,\ldots,h_{s-1}$–adic will be the only possible essential valuations of A which are not essential for $k[f,g_2]$. When we have done this we will be through, for it would follow by reduction that $A = k[f,g_{s+1}]$. Since k is algebraically closed, there is no loss in assuming $h_s = f$. Denote by A_f the localization of A at the prime fA. Then A_f is a rank one discrete valuation ring which has residue field transcendental over k. If A_f is centered on a minimal prime of $k[f,g_1]$, then it is essential for $k[f,g_1]$ and we take $g_1 = g_2$. Otherwise A_f is centered on a maximal ideal of $k[f,g]$, say $(f,g-\lambda)$. We replace g_1 by $g_1^1 = \frac{g_1-\lambda}{f}$. If A_f is not centered on a height one prime of $k[f, g_1^1]$, we repeat this to get g_1^2. There must exist an ℓ such that A_f is centered upon a height one prime of $k[f, g_1^\ell]$. This is because this selection procedure constitutes taking a sequence of quadratic transforms of $k[f,g_1]_{(f,g_1-\lambda)}$ along A_f. This sequence must reach A_f by (4.5). Set $g_2 = g_1^\ell$. Then A_f is an essential valuation of $k[f,g_2]$. Since the formation of $g_1^1,\ldots,g_1^\ell$ has taken place within $k[f,g_1,1/\varphi] = k[f,g_2,1/\varphi]$ and we know that the h_s–adic valuation is essential for the rings A and $k[f,g_2]$, we are reduced to $k[f,g_2] \subset A \subset k[f,g_2,h_1,\ldots,h_{s-1}]$. Thus $A = k[f,g_{s+1}]$ by reduction. □

(4.6) Remark. Professor Abhyankar has shown us how to remove the assumption that k is algebraically closed from (4.2). His argument uses the fact that for $\bar{k}$ algebraically closed, every $\bar{k}$–automorphism of $\bar{k}[X,Y]$ is a product of elementary automorphisms. We should also remark that (4.4) generalizes the results announced in the Notices, Nov. 1971, abstract no. 689–A27.

5. Problems.

The investigation of the cancellation problem provides numerous simply stated problems whose solutions appear to be non trivial. We take the opportunity to discuss a few of these in this closing section. Some of these questions are also posed in [AEH].

(5.1) Question. If A and B are domains and $A[X_1,\ldots,X_n] = B[Y_1,\ldots,Y_n]$, are the quotient fields of A and B isomorphic?

This is the "bi-rational cancellation problem" mentioned after (2.1).

(5.2) Question. If $A[X_1,\ldots,X_n] = B[Y_1,\ldots,Y_n]$ does there exist an isomorphism of A into B such that B is a finitely generated ring extension of the image of A?

This is easily seen to be the case for domains with n = 1.

In considering invariance and strong invariance when $A[X_1,\ldots,X_n] = B[Y_1,\ldots,Y_n]$, it would perhaps have been more precise to define n–invariant and n–strongly invariant.

(5.3) Question. Is it possible for a ring to be n-invariant (or n-strongly invariant) but not m invariant (m–

strongly invariant)?

(5.4) Question. If A is a strongly invariant integral domain and $A[X_1,\ldots,X_n] = B[Y_1,\ldots,Y_m]$ must it follow that $A \subset B$?

Of course we would like to know which ring-theoretic invariants are also invariants of the stable equivalence relation. While global and valuative dimensions offer no problems, Krull dimension does.

(5.5) Question. Suppose A is an integral domain of Krull dimension d and $A[X_1,\ldots,X_n] = B[Y_1,\ldots,Y_n]$. Does B have Krull dimension d?

We can offer one little result which may provide an approach to this problem

(5.6) Theorem. Let A be an integral domain of Krull dimension one and suppose $A[X] = B[Y]$ (one X; one Y), then B has Krull dimension one.

Proof. It is easy to see that an integral domain with a non zero Jacobson radical is strongly invariant. Thus we may assume that A has Jacobson radical zero. Thus every prime of A is an intersection of maximal ideals and it follows that A is a Hilbert ring. Since it is a homomorphic image of the Hilbert ring $A[X]$, B is a Hilbert ring. Let P be a depth two prime of B, and $P = \cap\, \mathfrak{m}_\alpha$ a representation of P as an intersection of maximal ideals of B. Then $P[Y] = \cap\, \mathfrak{m}_\alpha[Y]$ and each $\mathfrak{m}_\alpha[Y]$ is a depth one prime of $B[Y]$. If $P \neq 0$ then each $\mathfrak{m}_\alpha[Y]$ has height two and thus $\mathfrak{m}_\alpha[Y] \cap A = N_\alpha \neq 0$. This is because if $\mathfrak{m}_\alpha[Y] \cap A = 0$ then we could localize at the non-zero elements of A where $(\mathfrak{m}_\alpha[Y])^e$ could have height at most one. Therefore $N_\alpha[X] \subset \mathfrak{m}_\alpha[Y]$, and since depth $N_\alpha[X] = 1$, we must have $N_\alpha[X] = \mathfrak{m}_\alpha[Y]$ where N_α is a maximal ideal of A. Then we must have $(\cap\, N_\alpha)[X] = \cap(N_\alpha[X]) = \cap(\mathfrak{m}_\alpha[Y]) = P[Y]$. It then follows that $(\cap\, N_\alpha)$ is a non maximal prime of A, thus it must be zero. Therefore $P = 0$ and B is one dimensional. □

There is evidence to the effect that a Dedekind domain is either strongly invariant or a polynomial ring. If this were true, it would completely solve the cancellation problem for one dimensional noetherian domains. From [AEH] we have:

(5.7) Theorem. Let A be a Dedekind domain which is not strongly invariant, Then there is an element s in A such that $A[1/s]$ is the polynomials in one variable over some field k. If A contains the rational numbers then A is the polynomials in one variable over some field.

Using this result, the cancellation problem for Dedekind domains reduces to the following

(5.8) Question. Let V be a rank one discrete valuation ring with quotient field k and k(t) a simple transcendental extension of k. Let V* be an extension of V to k(t) such that $t \in V^*$ and the residue field of V* is algebraic over that of V. Let $A = V^* \cap k[t]$. Then A is a Dedekind domain [AEH]. Is A strongly invariant? Invariant?

The solution to this problem will finish the cancellation problem for Dedekind domains. We observe in [EH]

that a domain of the above type cannot be euclidian. This, of course solves the cancellation problem for euclidian domains. In this context it is also worth noting that if the valuation V^* above is an unramified extension of V if and only if resulting ring is a principal ideal domain. Since this construction is very easily implemented, it provides a large and quite tractable family of non euclidian principal ideal domains.

There is an interesting question on extending valuations which, if answered affirmatively would completely solve the cancellation problem for one dimensional noetherian domains by providing an affirmative answer to (5.8). It goes as follows:

(5.9) Question. Suppose V is a rank one discrete valuation ring with quotient field K. Let K(u,w) be a pure transcendental extension of degree two and V^* a rank one discrete extension of V to K(u,w). Suppose $Vu = V^* \cap K(u)$ and $Vw = V^* \cap K(w)$ have residue fields which are algebraic over the residue field of V. Must the residue field of V^* be algebraic over that of V?

If the residue field of V is of characteristic zero, or perfect, the answer is yes. To see this, extend V^* to a larger complete, rank one discrete valuation ring $\hat{V}^*$ whose coefficient field contains the algebraic closure of that of V. Now complete V, V_u and V_w within $\hat{V}^*$ to obtain $\hat{V}$, $\hat{V}_u$ and $\hat{V}_w$ respectively. Since the residue field of V is perfect we may assume that the coefficient fields of $\hat{V}$, $\hat{V}_u$ and $\hat{V}_w$ are all contained with that of $\hat{V}^*$ [N, (31.9), p.110]. Let W denote the completion of the integral closure of $\hat{V}$ in $\hat{V}^*$. Then the residue field of W is in fact the algebraic closure of that of V. Moreover $\hat{V}_u$ and $\hat{V}_w$ are integral over W. For if $f \in \hat{V}_u$, then since f is residually algebraic over $\hat{V}$, if $\mathfrak{m}$ is the maximal ideal of W, then $W[f]/\mathfrak{m}W[f]$ is a finite module over $W/\mathfrak{m}$. Hence W[f] is a finite W-module by [N, (30.6) p.103]. Now let W^* denote the integral closure of W in $\hat{V}^*$. Then W^* is an extension of both V_u and V_w with residue field algebraic over that of V. But $W^* \cap K(u,w)$ is V^*.

One can see from the above argument that an affirmative answer to (5.9) would follow from a negative answer to the following.

(5.10) Question. Suppose $V \subset V^*$ are complete rank one discrete valuation rings with residue fields k and k'(t). Suppose k' is an algebraic extension of k and t is transcendental over k. Can there exist distinct complete rank one discrete valuation rings V_u and V_w such that each has residue field k' and $V \subseteq V_u \cap V_w \subset V^*$?

In considering this question one may assume that k' is an infinite, purely inseparable extension of k, that the quotient field of V is algebraically closed in those of V_u and V_w, and that both V_u and V_w are algebraically closed in V^*.

Abhyankar and Moh have shown the following to be true.

(5.10) Let k be a field of characteristic zero and $f \in k[X,Y]$. If $k[X,Y]/(f) = k[t]$ then there exists $g \in k[X,Y]$ such that $k[X,Y] = k[f,g]$.

This raises the following possibility for solving the cancellation problem for k[X,Y]. Suppose we could affirmatively answer the following two questions.

(5.11) Question. Does (5.10) generalize to three variables? That is, suppose there exists $f \in k[X,Y,Z]$ such that $k[X,Y,Z]/(f) = k[u,v]$. Can one always find a g and an h such that $k[X,Y,Z] = k[f,g,h]$?

(5.12) Question. Suppose $A[T] = k[X,Y,Z]$, where A is a ring, T is transcendental over A and X,Y and Z are indeterminates. Does there exist a surjective homomorphism $\gamma: A \to k[t]$?

Were one to answer both of these questions affirmatively, then the kernel of γ would become the "f" in (4.1) and the cancellation problem would be solved for $k[X,Y]$, k a field of characteristic zero.

(5.13) Remark. Obviously, a problem similar to our is the following: **Suppose A and B are rings such that $A[[X]] \overset{\sigma}{\to} B[[Y]]$ is an isomorphism of rings of formal power series. Are A and B isomorphic?**

M. J. O'Malley has begun this investigation in [O'M]. He has proved the analogs of some of the results of [CE]. In particular he shows that with the above notation, if $\sigma(A) \subset B$ then $\sigma(A) = B$. He also shows that if the Jacobson radical of A is zero, then the above conditions imply $\sigma(A) = B$.

Acknowledgements: We are grateful to professor Abhyankar for the benefit of several stimulating conversations on this material. We also express our gratitude to Wanda Jones for her patience in typing the manuscript.

References

[A] S. Abhyankar, On the valuations centered on a local domain, Amer. J. Math. 78(1955) 321–348.

[AEH] S. Abhyankar, P. Eakin and W. Heinzer, On the uniqueness of the ring of coefficients in a polynomial ring, to appear J. Algebra.

[B] H. Bass, K–theory and stable algebra, IHES Publications Mathematiques 22(1964) 5–58.

[BR] J. Brewer and E. Rutter, Isomorphic polynomial rings, to appear Archiv der Math.

[C] C. Chevelley, Fundemental Concepts of Algebra, Academic Press, New York (1956).

[CE] D. Coleman and E. Enochs, Polynomial invariance of rings, Proc. AMS 27(1971) 247–262.

[EK] P. Eakin and K.K. Kubota, A note on the uniqueness of rings of coefficients in polynomial rings, Proc. AMS 23(1972) 333–341.

[EH] P. Eakin and W. Heinzer, Some Dedekind domains with specified class group and more non euclidian PID's, to appear.

[ES] P. Eakin and J. Silver, Rings which are almost polynomial rings, to appear Trans. AMS.

[H] M. Hochster, Non-uniqueness of the ring of coefficients in a polynomial ring, to appear Proc. AMS.

[K] I. Kaplansky, Modules over Dedekind rings and valuation rings, Trans. AMS 72(1952) 372–340.

[N] M. Nagata, Local Rings, Interscience, New York (1962).

[N,tvr] M. Nagata, A theorem on valuation rings and its applications, Nagoya Math J. 29(1967) 85–91.

[O'M] M. O'Malley, Power invariance of rings, to appear.

[R] C.P. Ramanujam, A topological characterization of the affine plane as an algebraic variety, Annals of Math. 94(1971) 69–88.

[V] W. Vasconcelos, On finitely generated flat modules, Trans. AMS 138(1969) 505–512.

THREE CONJECTURES ABOUT MODULES OVER POLYNOMIAL RINGS

by

David Eisenbud

and

E. Graham Evans, Jr.

I) Introduction. In [E-E,1] we observed that several results on generating modules and ideals, and several results about algebraic K-theory, all depended on essentially the same arguments. In [E-E,2] we showed that one of these results - the theorem of Kronecker that every ideal in a noetherian d-dimensional ring can be generated, up to radical, by d+1 elements- can be strengthened in case the ground ring is a polynomial ring : we showed that if R is a noetherian d-dimensional ring of the form S[x], then every ideal of R can be generated, up to radical, by d elements.

This result made us ask whether the other results discussed in [E-E, 1] could also be strengthened in case the ground ring is a polynomial ring. A brief look at the literature produced a number of examples of results of this type. We were led to conjecture that improvements in most of the theorems in [E-E,1] should be possible in the polynomial ring case.

In this paper we present our conjectures, with some evidence for their validity. In section 2 we state the conjectures and list a number of the known theorems which are special cases. In section 3, we prove that all our conjectures hold for polynomial rings over semi-local rings of positive dimension. We also establish a theorem on the number of generators of a projective module of rank 1 which is another special case of our conjectures.

We recall from [E-E,1] some of the terminology and results that we will use:

Let R be a ring and let M be a finitely generated R-module. Then $\mu(R,M)$ is the minimal number of generators of M over R.

If p is a prime ideal of R, then dim(p) is the Krull dimension of R/p. If M is an R-module, $m \in M$, and p a prime ideal of R, then m is basic in M at p if $\mu(R_p, M_p) > \mu(R_p, (M/Rm)_p)$. m is a basic element if it is basic in M at p for every prime ideal p of R.

The height of a prime ideal p of R, ht(p) is the Krull dimension of R_p. $\mathcal{P}_t$ is the set of prime ideals, p, of R such that $ht(p) \leq t$. If p is a prime ideal of R, $\dim_t(p)$ is the length of the longest chain of elements of $\mathcal{P}_t$ which contain p.

For the reader's convenience, we state two of the results from [E-E,1] that we will use in section 3. The first result concerns the existence of basic elements. We will state it only for the special case which we will use here.

Theorem A: [EE-1] : Let R be a noetherian ring, with $\dim R = d < \infty$. Let $M' \subseteq M$ be finitely generated R-modules, and suppose that for every prime ideal p of R, M' is $(\dim(p)+1)$-fold basic in M at p. If $m_1,\dots,m_u \in M'$ generate M', and if $r \in R$ is given such that $(r,m_1) \in R \oplus M$ is basic, then there is a basic element of M of the form m_1+rm', where $m' \in \sum_{i=2}^{u} Rm_i$.

The other result is that basic elements in projective modules are unimodular:

Lemma 1 [E-E,1]. If R is a commutative ring, and P is a finitely generated projective R-module, then an element $m \in P$ is basic if and only if it generates a free direct summand of P.

Section 2. The Conjectures, and a Survey of Known Special Cases.

The conjectures we wish to consider are the following:
Let S be a noetherian ring, and let $R = S[x]$ be the polynomial ring. Set $d = \dim R$.

1) If M is a finitely generated R-module such that $\mu(R_p,M_p) \geq d$ for every prime ideal p of R, then M has a basic element. In particular, if P is a projective R-module of rank d, then P has a free summand.

2) If P is a finitely generated projective R-module of rank $\geq d$ and if Q is a finitely generated projective R-module such that $Q \oplus P \cong Q \oplus P'$, then $P \cong P'$.

3) Let M be a finitely generated R-module, and let $\mathcal{P}$ be the set of primes of R such that $\dim(p) < d$. Set

$$n = \max_{p\in\mathcal{P}}(\mu(R_p,M_p) + \dim(p)).$$

Then M can be generated by n elements.

All three statements become true [E-E,1] if d is replaced by $d+1$.

It is easy to see that given their generality, these conjectures are the strongest possible. To see that this is so for conjecture 1, suppose that J is a maximal ideal of height d and let $M = J \oplus \dots \oplus J$ ($d-1$ times). Then M cannot have a basic element, for if $(r_1,\dots,r_{d-1})$ were a basic element, then J would be the radical of the ideal ΣRr_i contradicting Krull's principal ideal theorem.

To see that conjecture 2 fails if we replace d by $d-1$, let S be the coordinate ring of the real 2-sphere, $S = \mathbb{R}[x_1,x_2,x_3]/(x_1^2+x_2^2+x_3^2-1)$, and let $\bar{P}$ be the cokernel of $S \xrightarrow{(x_1,x_2,x_3)} S^3$. Then $\bar{P}$ is projective, and $\bar{P} \oplus S$ is free, but it is known that $\bar{P}$ is not free. If we set $R = S[x]$ and $P = \bar{P}\otimes_S R$ then $\operatorname{rank} P = 2 = \dim R-1$ and $R \oplus P \cong R^3$. However, if $P \cong R^2$, then $\bar{P} \cong P/XP \cong S^2$ which is a contradiction.

As for conjecture 3, Murthy [Mur,2] has given an example of an unmixed ideal I of height 2, in a ring of the form $R = K[X,Y,Z]$ where K is a field, such that I can be generated locally by 2 elements but requires 3 generators globally.

Conjecture 2 implies that if $R = K[X_1,\dots,X_d]$ with K a field, then every projective of rank $\geq d$ is free. This is a weak form of Serre's problem.

The following Corollary illustrates the application of conjecture 3:

<u>Corollary to Conjecture 3</u> : Let $R = S[X]$ be a noetherian polynomial ring, and let I be an ideal of R. Suppose that I can be generated locally by g elements. Then I can be generated by

$$\max(d , g+\dim R/I)$$

elements.

The Forster-Swan Theorem [E-E,1] implies in the above situation that I can be generated by

$$\max(d+1, g + \dim R/I)$$

elements.

It would be tempting to formulate a stronger version of conjecture 2, to parallel [E-E,1,Thm.Aiib].But this stronger form is false, as is any form strong enough to imply the Stable Range Theorem for the ring $K[X_1,\ldots,X_n]$, where K is the field of real numbers. (An example in [Vas] shows that $d+1$ is best possible value in this case).

We do not know whether, for a ring R satisfying this hypothesis of the conjectures, $E(d,R)$ is transitive on unimodular elements (See [Bass-2]).

We will now ennumerate the special cases of our conjectures that we have found in the literature.

The best known of these special cases is Seshadri's theorem that if S is a principal ideal ring, then every projective $S[X]$-module is free. This has been generalized by Serre, [Ser], Bass [Bass,1] and Murthy [Mur,1] to the case in which S is any 1-dimensional ring with only finitely many non-regular maximal ideals; in this case the theorem says that any projective $S[X]$-module is the direct sum of a free module and an ideal. (The freeness of all projectives in case

S is a principal ideal domain follows immediately from this statement.) This is precisely the conclusion of our conjecture 1 applied to projective modules. Moreover, in this case, conjecture 2 follows from conjecture 1 , since it is always possible to cancel a projective module from an ideal [Kap, p.76]. The truth of conjecture 3 in this case is open.

Another situation for which the truth of a part of conjecture 1, is known is that in which R is a polynomial ring in an odd number of variables over a field. Bass [Bass, 3] , Corollary 4.3, proves that in this case every projective P , with rank P = dim R , has a free summand.

On the non-projective side, the proof we gave in [E-E,2] for a slightly weaker result, actually implies that if , in conjecture 1 , M is a direct sum of ideals of R , then M has a basic element. (The connection between the result in [E-E,2] and basic elements is described in [E-E,1, Cor. 7 .]

We now turn to conjecture 2 . Endo [Endo] discussed the problem of when every projective over S[X,Y] is free, where S is a one-dimensional semi-local domain. His results give cases in which conjecture 2 holds. Bass and Schanuel [B-S] prove our conjecture for R = S[X] where S is a polynomial ring over a semi-local principal ideal domain. Bass in [Bass, 2] proves that if R = S[X] , where S is a polynomial ring over a semi-local ring of positive dimension, and if d = dim R , then E(n,R) is transitive on unimodular rows if $n > d$. This enables him to prove conjecture 2 in case P is free.

As for conjecture 3 , the statement is immediate in case dim S=0; it follows, for instance from the structure theorem for modules over a euclidean ring. Endo [Endo] proved conjecture 3 for maximal ideals in a ring of the form $S_o[X_1,\ldots,X_n]$, where S_o is a semilocal principal ideal domain. In [Ger], Geramita proves conjecture 3 for maximal ideals M of S[X] , where S is a Dedekind domain .

Davis and Geramita [D-G] prove it for maximal ideals over rings of the form $R = S_o[X_1,\ldots,X_n]$, where S_o is an arbitrary semilocal ring of positive dimension.

A particularly interesting special case of conjecture 3 is implied by a theorem of Murthy [Mur,2] , which shows that conjecture 3 holds if R is the ring of polynomials in 3 variables over a field, and M is any ideal of R .

Section 3 . 2 Special Cases of the Conjectures .

In this section we will establish all three of our conjectures in the case in which R has the form $R = S[X_1,\ldots,X_n]$, where S is a semi-local noetherian ring of positive dimension. This result includes a number of the known special cases of the conjectures which were mentioned in the previous section. We will also prove that conjecture 3 always holds for projective modules of rank 1 .

Theorem 1 . Let S be a semi-local ring with a noetherian spectrum of dimension > 0 , and let $R = S[X_1,\ldots,X_n]$. Set d = dim R . Then the three conjectures of section 2 hold for R .

Remarks. The hypothesis on S can be weakened to be that S has only finitely many prime ideals of maximal height. The only modification in the proof that would be necessary is the replacement of the Jacobson radical of S by the intersection of the primes of S of maximal height. A version of Theorem 1 using j-primes, etc., in the style of [E-E,1] may be proved just as we will prove Theorem 1 . Presumably, some non-commutative version of Theorem 1 , as in [E-E,1], is also true.

All three parts of Theorem 1 depend on the following lemma.

Lemma 2: Let R be a noetherian ring, and let I be an ideal of R . Let $K \subseteq M$ be finitely generated R-modules, and let t be an integer such that for every prime $p \in \theta_t$, K is $(\dim_t(p)+1)$-fold basic in M at p. Suppose that $r \in R$ and $k \in K$ are elements such that

a) $(r,k) \in R \oplus M$ is basic, and

b) The image of k in M/IM is basic.

Then there exists an element $k' \in IK$ such that $k + rk'$ is basic in M at p for all $p \in \theta_t$.

Proof of Theorem 1: Let J be the Jacobson radical of S, and set I = JR. The hypothesis on R and S shows that $\dim(R/I) < d$.

1) The second statement follows from the first because a basic element in a projective R-module generates a free summand [E-E,1, Lemma 1].

To prove the first statement, we note that, for every prime p of R which contains I, we have

$$\mu((R/I)_{\overline{p}}, (M/IM)_{\overline{p}}) = \mu(R_p, M_p) \geq d,$$

where we have written $\overline{p}$ for the reduction of p modulo I. Since $\dim(R/I) < d$, there exists a basic element in M/IM by [E-E, 1]. Let $m \in M$ be an element which reduces to this basic element. Applying the Lemma, with K = M and t = d-1, to the basic element $(1,m) \in R \oplus M$, we see that there exists $m' \in IM$ such that $m + m'$ is basic in M at every prime ideal $p \in \theta_{d-1}$.

We will show that $m + m'$ is basic at all primes of R. If q is a prime ideal of R with ht q = d, then $q \cap S$ is a maximal ideal of S. Thus $q \supseteq I$. On the other hand, $m + m'$ is basic modulo I since $m + m' \equiv m \pmod I$. This shows that $m + m'$ is basic at q, as required.

2.) We begin by making the familiar reduction to the case Q = R: there exists a projective module Q' such that $Q \oplus Q'$ is free, so it is enough to be able to cancel the rank 1 free summands of $Q \oplus Q'$ one at a time. Thus we may suppose $R \oplus P \cong R \oplus P'$. Let $f: R \oplus P' \to R \oplus P$ be the isomorphism. $(1,0) \epsilon R \oplus P'$ is clearly basic, so if $f((1,0)) = (r,m_o) \epsilon R \oplus P$, then (r,m_o) is basic. We will show that some automorphism a of $R \oplus P$ carries (r,m_o) to $(1,0)$. We will thus obtain a commutative diagram with exact rows of the following form:

$$\begin{array}{ccccccccc} 0 & \to & R & \to & R \oplus P' & \to & P' & \to & 0 \\ & & \| & & \downarrow{\scriptstyle af} & & & & \\ 0 & \to & R & \to & R \oplus P & \to & P & \to & 0 \end{array} .$$

This shows that $P \cong P.'$

To construct a, we again use [E-E, 1 Theorem A] and the fact that $\dim (R/I) < d$ to show that there exists an element m_1 in P such that $\overline{m}_o + r\overline{m}_1$ is basic in P/IP, where , we have written $-$ to denote reduction modulo I.

We can now apply the Lemma with $K = M$ and $t = d-1$ to the element $(r, m_o + rm_1) \epsilon R \oplus P$. By the Lemma, there exists an element $m_2 \epsilon IP$ such that $m_o + rm_1 + rm_2$ is basic at all primes $p \epsilon \mathcal{P}_{d-1}$. As in part 1, above, it follows that the element $m = m_o + rm_1 + rm_2$ is basic in P.

The rest of the proof follows the pattern in [Bass, 2] and [E-E, 1,Cor.4]: If α denotes the map $R \to P$ carrying 1 to $m_1 + m_2$, β denotes a map $P \to R$ carrying m to $1-r$ (such maps exist because, by Lemma 1, m generates a free summand of P), and γ denotes the map $R \to P$ taking 1 to $-m$, then we may take

$$a = \begin{pmatrix} 1 + \beta\alpha & \beta \\ \gamma + \alpha + \gamma\beta\alpha & 1 + \gamma\beta \end{pmatrix}$$

3) Unless $n < \max\limits_{\substack{\dim p = d \\ M_p \neq 0}} (d + \mu(R_p, M_p))$, this is the conclusion of the usual Forster-Swan theorem [E-E,1]. Therefore there must exist primes p of dimension d such that $M_p \neq 0$, and so, a fortiori, there exist primes p of dimension $< d-1$ such that $M_p \neq 0$. Thus $n \geq d$.

Now suppose $\mu(R,M) = t > n$. It follows that there is a short exact sequence of the form

(*) $\quad 0 \to K \to R^t \to M \to 0$

Following the pattern of [E-E,1, Cor.5], we will show that there exists an element $m \in K$ which is basic in R^t, so that $R^t = Rm \oplus P$, by [E-E,1 , Lemma 1] , for some projective module P. We will then have rank $P = t-1 \geq d$, so by part 2), above, P is free of rank $t-1$. On the other hand, the epimorphism $R^t \to M$ induces an epimorphism $P \to M$, so M can be generated by $t-1$ elements. This contradicts our assumption that $\mu(R,M) = t$, proving the theorem.

It remains to show that K contains a basic element of R^t. From (*) and the assumption $t > n$, it follows that K is ($\dim p+1$)-fold basic in R^t at every prime p such that $\dim p < d$, and K is d-fold basic in R^t at primes p such that $\dim P = d$.

It is easy to see that if a prime ideal p contains the ideal I, then the image of K in R^t/IR^t is just as basic at p as is K in R^t. Thus [E-E, 1, Thm. A] implies that the image of K in R^t/IR^t contains a basic element $\bar{k}$ of R^t/IR^t. Let $k \in K$ be an element which reduces to $\bar{k}$ modulo IR^t. Applying the Lemma with $t = d-1$ to the basic element $(1,k) \in R \oplus R^t$, we see that there exists an element $k' \in IK$ such that $k + k'$ is basic at all prime ideals $p \in \theta_{d-1}$. Since k is basic in R^t modulo I, it follows as in the proof of 1) that $k+k' = m$ is basic in R^t, as required.

<u>Sketch of Proof of Lemma 2</u> : The proof of this Lemma follows the pattern of the proof of Theorem A given in [E-E, 1] so closely that we will not give it in detail. Instead, we will remark on the points in the proof of Theorem A at which changes must be made. We assume that the reader has a copy of [E-E,1] before him.

The main difference between Lemma 2 of this paper, and Theorem A ii) b) of [E-E,1] with $A = R$, is that in Lemma 2, m_1 is assumed basic mod I , and we wish to produce elements a_i (in the notation of Theorem A!) which lie in I . (The appearance of the sets θ_t is an essentially trivial change). The elements a_i are actually obtained by a number of applications of Lemma 3 of [E-E,1]. In the notation of Lemma 3 , it suffices to prove that we may take $a_1 \epsilon I$ (the other elements a_j in Lemma 3 don't matter). Turning to the proof of Lemma 3 , in section 5 of [E-E,1] , we see that a_1 is chosen so that, in the notation of [E-E,1],

$$m_1 + aa_1m_1$$

is basic in M at a certain finite list of primes $p_1,\dots,p_v$. Suppose that m_1 is basic mod I, and that $p_1,\dots,p_{v'}$ are the primes of this list that contain I . Then any choice of $a_1 \epsilon I$ makes $m_1+aa_1m_1$ basic mod I, and therefore basic at each of the primes $p_1,\dots,p_{v'}$. Thus it suffices to deal with the primes $p_{v'+1},\dots,p_v$, which do not contain I . To do this, we first choose a_1 in the way that the proof of Lemma 3 instructs us- not necessarily in I. Then we pick $s \epsilon I$ such that $s \notin \bigcup_{i=v'+1}^{v} P_i$; this is possible, since otherwise I would be contained in one of the primes in the union. We can now replace a_1 by $sa_1 \epsilon I$ and continue with the proof as given in [E-E,1].

The next theorem covers a special case of conjecture 3). The proof is similar to that in [E-E,2].

Theorem 2. Let S be a noetherian ring, and let R = S[X] be the polynomial ring, and let P be a projective R-module of rank 1 . If dim R = d , then P can be generated by d elements.

Proof. We proceed by induction on d-1 = dim S . If dim S = 0 , then S is artinian. Let N be its nilpotent radical. By Nakayama's lemma,

$$\mu(R,P) = \mu(R/NR, P/NP).$$

Thus we may assume N = 0, so that S is a direct product of fields. But in this case R is a direct product of principal ideal domains, so P is cyclic.

Now suppose that $\dim S > 0$. Let $\mathcal{U} \subseteq S$ be the multiplicative subset which is the complement of the union of the minimal primes of S .

Then $S_{\mathcal{U}}$ has dimension 0 , so by induction, $P_{\mathcal{U}}$ can be generated over $S_{\mathcal{U}}[X] = R$ by one element $p_1 \in P$. It follows that there exists an element $u \in \mathcal{U}$ such that

$$(3) \qquad uP \subseteq Rp_1$$

Since u is not in any of the minimal primes of S , $\dim S/(u) < \dim S$. Thus, by induction, the rank 1 projective $R/(u)$-module P/uP can be generated by d-1 element $\overline{p}_2,\ldots,\overline{p}_d$, that is ,

$$(4) \qquad P/uP = \sum_{i=2}^{d} R\overline{p}_i .$$

We claim that if $p_2,\ldots,p_d$ are any elements of P which reduce modulo (u) to $\overline{p}_2,\ldots,\overline{p}_d$, then

$$P = \sum_{i=1}^{d} Rp_i$$

It is enough to prove that this holds after localizing at an arbitrary prime ideal q of R . If $u \notin q$ then

$$(5) \qquad P_q = \Big(\sum_{i=1}^{d} Rp_i \Big)_q$$

follows immediately from (3). If, on the other hand, $u \in q$, then (5) follows from Nakayama's lemma and (4). This concludes the proof.

REFERENCES

Bass, H., [Bass, 1] "Torsion Free and Projective Modules," Trans. A.M.S. 102 (1962), 319-327.

[Bass, 2] "K-Theory and Stable Algebra " Publ. Math. IHES no. 22 (1964), 5-60.

[Bass, 3] "Modules Which Support Nonsingular Forms," J. Alg. 13 (1969), 246-252.

Bass, H., and Schanuel, S. [B-S] "The Homotopy Theory of Projective Modules," Bull. A.M.S. 68 (1962), 425-428.

Davis, E., and Geramita, A., [D-G] "Maximal Ideals in Polynomial Rings." these Proceedings.

Eisenbud, D., and Evans, E. [E-E,1] "Generating Modules Efficiently: Theorems from Algebraic K-Theory." J. Alg., To appear.

[E-E, 2] "Every Algebraic Set in n-space is the Intersection of n Hypersurfaces." To appear.

Endo, S. [Endo] "Projective Modules Over Polynomial Rings." J. Math Soc. Japan 15 (1963), 339- 352.

Geramita, A. [Ger], "Maximal Ideals in Polynomial Rings ," Queen's University Mathematical Preprint no. 1971-56.

Kaplansky, I., [Kap] Infinite Abelian Groups, Revised edition, The University of Michigan, Ann Arbor, (1969).

Kronecker, L., [Kro] "Grundzüge eine arithmetischen Theorie der algebraischen Grossen," J. Reine. Angew. Math. 92(1882), 1-123.

Murthy, P. [Mur,1] "Projective Modules Over a Class of Polynomial Rings," Math. Z. 88 (1965), 184-189.

[Mur, 2] "Generators for Certain Ideals in Regular Rings of Dimension three" To appear.

Serre, J.-P. [Ser] "Sur les Modules Projectifs," Sem. Dubriel-Pisot. t. 14 , 1960-1961, no. 2 .

Seshadri, C. [Ses] "Triviality of Vector Bundles over the Affine Space K^2," Proc. Nat. Acad. Sci. 44(1958), 456-458.

Vaserstein, L. [Vas] "Stable Rank of Rings and Dimensionality of Topological Spaces," Functional Analysis and its Applications 5 , (1971), 102-110.

Brandeis University
Waltham, Massachusetts

M.I.T.
Cambridge, Massachusetts

PRÜFER-LIKE CONDITIONS ON THE SET OF OVERRINGS OF AN INTEGRAL DOMAIN

Robert Gilmer, Department of Mathematics
Florida State University, Tallahassee, Florida, 32306

This paper considers ten conditions on the set of overrings of an integral domain D with identity. Each of these conditions is satisfied if D is a Prüfer domain. Relations among the conditions are discussed, and several related questions are mentioned.

Let D be an integral domain with identity with quotient field K. By an overring of D, we mean a subring of K containing D. We denote by Σ the set of overrings of D, and we consider the following eleven conditions.

(0) D is a Prüfer domain; that is, finitely generated nonzero ideals of D are invertible.

(1) Each valuation overring of D (that is, an overring of D that is a valuation ring) is a quotient ring of D.

(2) Each overring of D is a Prüfer domain.

(3) Each overring of D is integrally closed.

(4) Each overring of D is flat as a D-module.

(5) Each overring of D is an intersection of quotient rings of D.

(6) If $J \in \Sigma$, then the prime ideals of J are extensions of prime ideals of D.

(7) If $J \in \Sigma$ and if P is a prime ideal of D, then there is at most one prime ideal of J lying over P.

(8) If $J \in \Sigma$ and if P is a prime ideal of D, then a chain of prime ideals of J lying over P has at most one member.

(9) The set Σ is closed under addition—that is, if $J_1, J_2 \in \Sigma$, then $J_1 + J_2 = \{x_1 + x_2 \mid x_i \in J_i\}$ is in Σ.

(10) $\dim J \leq \dim D$ for each J in Σ.

In the sequel we discuss relations among the preceding conditions. In particular, conditions (0) - (4) are equivalent and they imply conditions (5) - (10); if D is integrally closed, then conditions (0) - (9) are equivalent.

Conditions (1) - (3) and Some Related Questions

W. Krull in [42, p. 554] established the equivalence of (0) and (1)—that is, D is a Prüfer domain[1] if and only if each valuation overring of D is a quotient ring of D. It then follows easily that (0) and (2) are equivalent, for if $J, V \in \Sigma$ with $J \subseteq V$, and if V is a quotient ring of D, then V is also a quotient ring of J.

Since a Prüfer domain is integrally closed, the implication (0) → (3) follows from the equivalence of (0) and (2). E. D. Davis proved the reverse implication in [12, p. 198]; he also extended his results to commutative rings R with identity with few zero divisors, the definition of such a ring being that the set of zero divisors of R is a finite union of prime ideals of R.

Several questions related to the equivalence of (0) and (3) have been considered in the literature. A dual question that is easy to answer is: Under what conditions is each subring of D with identity integrally closed? The answer to this question is contained in [20, Theorem 1].

THEOREM A. The following conditions are equivalent.

(1) Each subring of D with identity is integrally closed.

(2) D is an algebraic extension field of a finite field or D has quotient field Q, the field of rational numbers.

---

[1]Krull uses the term "Multiplikation ring" in [42] instead of "Prüfer domain". As Krull noted, his use of this term was in conflict with its meaning in older literature. Current usage of the term multiplication ring has reverted to that of the older literature—a commutative ring R in which the containment relation $A \subseteq B$ for ideals implies the existence of an ideal C such that $A = BC$ (that is, "every divisor is a factor"); see [27] for more on multiplication rings. The first use of the term "Prüfer domain" (or rather, "Prüfer ring") that I have found in the literature appears in H. Cartan and S. Eilenberg [11, p. 133]; I. Kaplansky repeats the terminology in [40]. The term "Prüfer ring", as it is currently used, includes commutative rings with zero divisors; see [32] and [46, Chapter 10].

It follows from Theorem A that if each subring J of D with identity is integrally closed, then each such J is a Euclidean domain with the property that J/A is finite for each nonzero ideal A of J.

In [19, Theorems 3, 4], R. Gilmer gives necessary and sufficient conditions, reminiscent of those in (2) of Theorem A, in order that each subring of D should be Noetherian; W. Borho [8] has investigated this question for rings (not necessarily commutative) with zero divisors, and the "Noetherian pairs" of A. Wadsworth [54] are another variant of the same question[2].

In [27], Gilmer and J. Mott considered several questions of the following type: Determine necessary and sufficient conditions in order that each integrally closed subring of D should have property P. To give an indication of the results of [27], we cite a portion of its Theorem 3.1.

THEOREM B. The following conditions are equivalent.

(1) Each integrally closed subring of D is completely integrally closed.

(2) Either K has characteristic 0 and K/Q is algebraic, or K has characteristic $p \neq 0$ and tr.d. $K/GF(p) \leq 1$.

(3) Each integrally closed subring of K is a Prüfer domain.

In analogy with Noetherian pairs, we could, of course, define an integrally closed pair to be a pair (R, S) of commutative rings with a common identity such that R is a subring of S and each subring of S containing R is integrally closed[3].

Flat Overrings

In [50], F. Richman proved that D is a Prüfer domain if and only if each overring of D is flat as a D-module. This particular equivalence is useful in relating the structure of D to that of an overring J of D, primarily because of Theorem C,

[2]For more on questions of the form "characterize rings R such that each subring of R has property P", see [19, Theorem 5], [25], and [30].

[3]Several persons, including E. D. Davis, A. Grams, I. Kaplansky, T. Parker, and A. Wadsworth, have shared with me some observations about such pairs after the conclusion of my talk; see, for example, the next paper (by Davis) in these Proceedings.

which we cite later. At this point, we remark that the equivalence of (0), (3), and (4) generalizes to the case of rings with zero divisors; see [32] and [46, Theorem 10.18].

There are several papers in the literature that touch on the subject of flat overrings. The papers [1, 2, 3] of T. Akiba dwell on the topic, while [50], [45], [35], [32], and [5] also contain some of the general theory of flat overrings. As a nice summary statement, we cite a result that appeared in a preprint of [5].

THEOREM C. Let R be a commutative ring with identity with total quotient ring T, and let R' be a subring of T containing R such that R' is flat as an R-module.

(a) There is a generalized multiplicative system[4] S in R such that $R' = R_S$ and $AR' = R'$ for each A in S.

Let A and B be ideals of R, and let A' be an ideal of R'.

(b) $A_S = AR'$.

(c) $AR' = R'$ if and only if there exists $B \in S$ such that $A \supseteq B$.

(d) Let Q be a P-primary ideal of R such that $QR' \subset R'$. Then $PR' \subset R'$, PR' is prime in R', QR' is PR'-primary, $P = PR' \cap R$, and $Q = QR' \cap R$.

(e) $A' = (A' \cap R)R'$.

(f) If $A' = (a_1', \ldots, a_m')$ is finitely generated, then there exist $b_1, \ldots, b_n \in R$ such that for $1 \le i \le m$, $1 \le j \le n$, $a_i'b_j \in R$ and $A' = \sum_{i=1}^{m} \sum_{j=1}^{n} a_i'b_jR'$.

---

[4] By a <u>generalized multiplicative system</u> in a commutative ring T, we mean a nonempty family $S = \{I_\alpha\}$ of nonempty subsets of T (in forming generalized quotient rings, there is no loss of generality in assuming that each I_α is an ideal of T) such that $I_\alpha I_\beta = \{\sum_1^n x_iy_i \mid x_i \in I_\alpha, y_i \in I_\beta\} \in S$ for all α and β. By the <u>generalized quotient ring</u> T_S <u>of</u> T <u>with respect to</u> S, we mean the set of elements x in the total quotient ring of T such that $xI_\alpha \subseteq T$ for some I_α in S. The notion of a generalized quotient ring, which obviously generalizes the concept of a quotient ring with respect to a regular multiplicative system, seems to have originated with W. Krull [44, Section 2]. More recently, D. Kirby [41], L. Budach [9], M. Griffin [32], W. Heinzer, J. Ohm, and R. Pendleton [35], J. Arnold and J. Brewer [5], and H. S. Butts and C. Spaht [10] have used the concept. It is interesting to note that the set of generalized quotient rings of D includes the set of overrings J of D that can be written as an intersection of localizations of D [35, Prop. 4.3 and Cor. 4.4].

(g) $(A \cap B)R' = AR' \cap BR'$.

(h) $\sqrt{(AR')} = (\sqrt{A})R'$.

(i) If B is finitely generated, then $[AR':BR'] = [A:B]R'$.

Of several homological characterizations of Prüfer domains, we mention one, in terms of overrings, here. H. Storrer [53] has proved that D is a Prüfer domain if and only if the injection map of D into J is an epimorphism in the category of commutative rings for each J in Σ. Storrer also proved that D is a Prüfer domain if and only if the D-module $D[s] \otimes_D D[s]$ is torsion-free for each s in K[5].

Conditions (5) - (8) and Some Related Results

As we have previously observed, a flat overring of D is an intersection of localizations of D. Moreover, part (e) of Theorem C shows that condition (5) is satisfied for each flat overring J of D. It is clear that (6) implies (7) and that (7) implies (8). Consequently, conditions (5) - (8) are satisfied if D is a Prüfer domain.

E. D. Davis asked in [12, p. 200] if a domain D for which each overring is an intersection of localizations of D (and hence each overring of D is a generalized quotient ring of D) is, in fact, a Prüfer domain. He proved that the answer to this question is affirmative if each prime ideal of D has finite height. In [21], Gilmer and Heinzer conducted a thorough investigation into the theory of a domain D for which each overring is an intersection of localizations; they called such a domain *a QQR-domain* or *a domain with the QQR-property*. Gilmer and Heinzer proved that the integral closure of a QQR-domain is a Prüfer domain; moreover, if each valuation overring of D is an intersection of localizations of D (that is, D is *a VQR-domain*) and if the ascending chain condition for prime ideals holds in D, then D is a Prüfer domain. Finally, [21] contains examples that show that a VQR-domain need not have the QQR-property, and that a QQR-domain need not be a Prüfer domain.

---

[5] A. Hattori [33] proved that D is a Prüfer domain if and only if for all torsion-free D-modules M and N, $M \otimes_D N$ is torsion-free; some of Hattori's results were generalized by S. Endo in [13].

It seems appropriate at this point to mention the construction used by Gilmer and Heinzer in [21] to obtain their examples. If V is a valuation ring with maximal ideal M, if ϕ is the canonical homomorphism of V onto $\Delta = V/M$, and if D is a subring of Δ, then $D_1 = \phi^{-1}(D)$ is a subring of V containing M. The structure of D_1 reflects both the structure of V and the structure of Δ as a ring extension of D, and hence a rather wide range of properties can be realized by such domains D_1. Because the domains D_1 have usually been used in the literature to provide examples, one frequently assumes that $V = \Delta + M$—that is, the residue field of V is realized as a subring of V; in this case, $D_1 = D + M$, and for this reason, I refer to such a construction in Appendix 2 of [16] as the <u>$D + M$ construction</u>. In the form just described, the earliest use of the $D + M$ construction that I have found in the literature is in a paper [43, p. 670] by Krull. Krull gives there an example of a one-dimensional quasi-local integrally closed domain J that is not a valuation ring. His description of J is the following: Let k be a field and let J be the set of all rational functions $f(X, Y)/g(X, Y)$ in two variables over k such that $g(0, Y) \neq 0$ and $f(0, Y)/g(0, Y) \in k$. In terms of a $D + M$ construction, Krull's example amounts to taking $J = k + M$, where M is the maximal ideal of the rank one discrete valuation ring $k[X, Y]_{(X)} = k(Y) + M$. If we consider the closely related construction $R + B$, where R is a subring of the commutative ring S and B is an ideal of the polynomial ring $S[\{X_\lambda\}]$ contained in $(\{X_\lambda\})$, then such a construction can be found in the classical paper [49, p. 19] of H. Prüfer[6]. To obtain an example of a domain with property $L\Delta$ that does not have property $B\Delta$ (in modern terminology, this is an example of an integrally closed domain that is not completely integrally closed), Prüfer gave the example $D + (X, Y)K[X, Y]$, where D is integrally closed and $D \neq K$. In [42, p. 569], Krull uses the same domain as an example of an integrally closed domain on which the v-operation is not endlich arithmetisch brauchbar.

[6]Prüfer's paper [49] represents the first study of Prufer domains, <u>per se</u>, in the literature. With proper modesty, Prüfer's term for an integral domain with identity in which nonzero finitely generated ideals are invertible was <u>a domain with property LB</u>.

Another important use of the D + M construction is in the paper [52, p. 604] of A. Seidenberg; Seidenberg used such domains to prove that if n and k are positive integers such that $n + 1 \leq k \leq 2n + 1$, then there is an integrally closed domain J such that dim J = n and dim J[X] = k. In recent years, the D + M construction and related construction for polynomial rings have appeared frequently in the literature[7], and some recent papers have contained theorems concerning the structure of such constructions; see [29], [22], [16, Appendix 2], [7]. In [18], Gilmer has generalized some of these theorems to finite intersections $\cap_{i=1}^{n} (D_i + M_i)$.

We return to our discussion of relations between conditions (0), (5) - (8). One of the examples of Gilmer and Heinzer in [21] shows that conditions (5), (6), and (9) do not imply (0). On the other hand, conditions (7) and (8) are equivalent and are, in fact, equivalent to the condition that the integral closure of D is Prüfer [16, Th. 16.10 and Th. 22.2]. Consequently, conditions (0) - (8) are equivalent for an integrally closed domain.

If each overring of D is a quotient ring of D, then it follows from the equivalence of (0) and (2) that D is a Prüfer domain. Such domains were considered independently by Davis [12] and by Gilmer and Ohm [28]. Following [28], we say that D has the <u>QR-property</u> if each overring of D is a quotient ring of D. A Prüfer domain need not have the QR-property. In fact, the following theorem was proved independently by Davis, Gilmer and Ohm, and by O. Goldman [31].

THEOREM D. If D is Noetherian, then D has the QR-property if and only if D is a Dedekind domain with torsion class group.

In [28], Gilmer and Ohm prove that a Prüfer domain D has the QR-property if a power of each finitely generated ideal of D is principal; in [48], Pendeleton strengthened this result to: The Prüfer domain D has the QR-property if and only if for each finitely generated ideal A of D, $\sqrt{A}$ is the radical of a principal ideal of D. Must a domain with the QR-property have torsion class group? This

---

[7] In fact, I have heard the statement made in jest that if an example exists, it can be realized in the form D + M. Since each domain with identity is trivially of the form, the statement is literally true, but in making this observation, I avoid the spirit of the speaker.

question remained open for several years; it was answered in the negative by Heinzer in [34]. The example that Heinzer gave (not a $D + M$ construction) is related to a construction of Goldman in [31].

The search for a characterization of Prüfer domains in terms of overrings being quotient rings is not lost, however. Both Arnold and Brewer [5] and Butts and Spaht [10] have obtained such a characterization. Specifically, Theorem 1.5 of [5] states that D is a Prüfer domain if and only if each overring of D is of the form D_S for some generalized multiplicative system $S = \{I_\alpha\}$ of invertible ideals of D.

Arithmetic Relations on the Set of Overrings of D

In the terminology of [24], D is a Δ-domain if for all $J_1, J_2 \in \Sigma$, $J_1 + J_2 = \{a_1 + a_2 \mid a_i \in J_i\}$ is a subring of K. A Prüfer domain is a Δ-domain; this follows since a valuation ring is obviously a Δ-domain and since a domain D is a Δ-domain if and only if D_M is a Δ-domain for each maximal ideal M of D. More generally, a domain with the QQR-property is a Δ-domain [24, Theorem 5], and hence a Δ-domain need not be a Prüfer domain. It is true, however, that the integral closure of a Δ-domain is a Prüfer domain.

Condition (9) is distinguished from the other conditions on our list because it is primarily arithmetic in nature. In fact, it is true that D is a Δ-domain if and only if $xy \in D[x] + D[y]$ for all $x, y \in K$. Several characterizations of Prüfer domains, in terms of arithmetic relations on the set of ideals, are known. As a fair summary of these results, we state Theorem E (see [38] and [16, Section 21]).

THEOREM E. The following conditions are equivalent.

(a) D is a Prüfer domain.

(b) $A \cap (B + C) = (A \cap B) + (A \cap C)$ for all ideals A, B, C of D.

(c) $(A + B)(A \cap B) = AB$ for all ideals A, B of D.

(d) $A(B \cap C) = AB \cap AC$ for all ideals A, B, C of D.

(e) $A:B + B:A = D$ for all A, B finitely generated ideals of D.

(f) $(A + B):C = A:C + B:C$ for all ideals A, B, C of D with C finitely generated.

(g) $C:(A \cap B) = C:A + C:B$ for all ideals A, B, C of D with A and B

finitely generated.

(h) $AB = AC$, for ideals A, B, C of D with A finitely generated and nonzero, implies that $B = C$.

Several of the conditions of Theorem E have been considered for commutative rings R with identity [46, pp. 150,151]. This is especially true of condition (b) —the condition that the lattice of ideals of R, under $+$ and $\cap$, should be distributive. L. Fuchs in [14] refers to such rings as arithmetischer Ringe, and C. U. Jensen has investigated these rings (arithmetical rings, in Jensen's terminology) in a series of papers (see, in particular, [39]).

Besides [24], a few other papers that have touched on the question of characterizing Prüfer domains D in terms of arithmetic relations on the set of overrings of D are [47], [17], and [23].

Dimension Theory of Overrings

It follows from part (e) of Theorem C that $\dim J \leq \dim D$ if J is a flat overring of D. Thus condition (10) is satisfied if D is a Prüfer domain; more generally, condition (10) follows from condition (8). Since dimension is preserved by integral extensions, it is clear that condition (10) is satisfied for D if and only if it is satisfied for D', the integral closure of D. But unlike the other conditions we have considered, an integrally closed domain satisfying (10) need not be a Prüfer domain. For finite dimensional domains, the following theorem gives a characterization of domains satisfying (10).

THEOREM F. If D has finite dimension n, then the following conditions are equivalent.

(a) $\dim J \leq n$ for each J in Σ.

(b) The valuative dimension of D, $\dim_v D$, is n.

(c) $\dim D[X_1, \ldots, X_n] = 2n$.

A proof of Theorem F is given in [16, pp. 346-9]; Arnold proves a generalization of Theorem F in Theorem 6 of [4]. The notion of valuative dimension was introduced by P. Jaffard in [36] (see also [37, Chapitre IV]); $\dim_v D$ is defined to be sup {dim V | V is a valuation overring of D}. The inequality $\dim D \le \dim_v D$ always holds, and it follows from Theorem F that $\dim D = \dim_v D$ if D is a Prüfer domain or if D is Noetherian. In particular, condition (10) is satisfied for each finite-dimensional integrally closed Noetherian domain D_1, but D_1 is a Prüfer domain if and only if $\dim D_1$ is at most 1. There is one nontrivial case in which condition (10) implies that the integral closure of D is a Prüfer domain—the case where $\dim D = 1$. This result—that $\dim_v D = 1$ implies that the integral closure of D is a Prüfer domain—was proved by Seidenberg in [51, p. 511]; see also [15].

If k is a positive integer and if m is ∞ or is a positive integer greater than or equal to k, then there exists an integrally closed domain D_k such that $\dim D_k = k$ and $\dim_v D_k = m$. Such a domain D_k can be realized by a D + M construction [16, p. 572]. In general, the sequence {dim D, dim $D[X_1]$, dim $D[X_1, X_2]$, ...} can be wild indeed; for example, see [7, Section 5] and [6]. But if $\dim_v D = r < \infty$, then for $s \ge r$, $\dim D[X_1, \ldots, X_s] = r + s$ [36, Théorème 3].

Resume

In summary, conditions (0) - (4) are equivalent, and these conditions imply conditions (5) - (10). The integral closure of a domain satisfying one of the conditions (5) - (9) is a Prüfer domain, but the domain of Example 4.1 of [21] shows that the combination of conditions (5) - (10) does not imply that D is integrally closed. Finally, for each positive integer $n > 1$, there is an n-dimensional integrally closed domain D_n such that condition (10) is satisfied for D_n, but D_n is not a Prüfer domain.

REFERENCES

[1] Akiba, T. Remarks on generalized quotient rings, Proc. Japan Acad. 40, 801-806 (1964).

[2] ________ Remarks on generalized rings of quotients II, J. Math. Kyoto Univ. 5, 39-44 (1965).

[3] ________ Remarks on generalized rings of quotients III, J. Math. Kyoto Univ. 9, 205-212 (1969).

[4] Arnold, J. T. On the dimension theory of overrings of an integral domain, Trans. Amer. Math. Soc. 138, 313-326 (1969).

[5] Arnold, J. T., and Brewer, J. W. On flat overrings, ideal transforms, and generalized transforms of a commutative ring, J. Algebra 18, 254-263 (1971).

[6] Arnold, J. T., and Gilmer, R. The dimension sequence of a commutative ring, in preparation.

[7] Bastida, E., and Gilmer, R. Overrings and divisorial ideals of rings of the form D + M, preprint.

[8] Borho, W. Die torsionsfreien Ringe mit lauter Noetherschen Unterringen, preprint.

[9] Budach, L. Quotientenfunktoren und Erweiterungstheorie, Math. Forschungsberichte 22, VEB Deutscher Verlag der Wiss. Berlin, 1967.

[10] Butts, H. S., and Spaht, C. G. Generalized quotient rings, to appear in Math. Nach.

[11] Cartan, H., and Eilenberg, S. Homological Algebra, Princeton Univ. Press, Princeton, N. J. (1956).

[12] Davis, E. D. Overrings of commutative rings. II. Integrally closed overrings, Trans. Amer. Math. Soc. 110, 196-212 (1964).

[13] Endo, S. On semi-hereditary rings, J. Math. Soc. Japan 13, 109-119 (1961).

[14] Fuchs, L. Über die Ideale arithmetischer Ringe, Comment. Math. Helv. 23, 334-341 (1949).

[15] Gilmer, R. Domains in which valuation ideals are prime powers, Arch. Math. 17, 210-215 (1966).

[16] _______ Multiplicative Ideal Theory, Queen's University, Kingston, Ontario (1968).

[17] _______ On a condition of J. Ohm for integral domains, Canad. J. Math. 20, 970-983 (1968).

[18] _______ Two constructions of Prüfer domains, J. Reine Angew. Math. 239/240, 153-162 (1969).

[19] _______ Integral domains with Noetherian subrings, Comment. Math. Helv. 45, 129-134 (1970).

[20] _______ Domains with integrally closed subrings, Math. Jap. 16, 9-11 (1971).

[21] Gilmer, R., and Heinzer, W. Intersections of quotient rings of an integral domain, J. Math. Kyoto Univ. 7, 133-150 (1967).

[22] ______ On the number of generators of an invertible ideal, J. Algebra 14, 139-151 (1970).

[23] Gilmer, R., and Huckaba, J. A. The transform formula for ideals, J. Algebra 21, 191-215 (1972).

[24] ______ Δ-rings, preprint.

[25] Gilmer, R., Lea, R., and O'Malley, M. Rings whose proper subrings have property P, to appear in Acta Sci. Math. (Szeged).

[26] Gilmer, R., and Mott, J. L. Multiplication rings as rings in which ideals with prime radical are primary, Trans. Amer. Math. Soc. 114, 40-52 (1965).

[27] ______ Integrally closed subrings of an integral domain, Trans. Amer. Math. Soc. 154, 239-250 (1971).

[28] Gilmer, R., and Ohm, J. Integral domains with quotient overrings, Math. Ann. 53, 97-103 (1964).

[29] ______ Primary ideals and valuation ideals, Trans. Amer. Math. Soc. 117, 237-250 (1965).

[30] Gilmer, R., and O'Malley, M. Non-Noetherian rings for which each proper subring is Noetherian, to appear in Math. Scand.

[31] Goldman, O. On a special class of Dedekind domains, Topology 3, 113-118 (1964).

[32] Griffin, M. Prüfer rings with zero divisors, J. Reine Angew. Math. 239/240, 55-67 (1969).

[33] Hattori, A. On Prüfer rings, J. Math. Soc. Japan 9, 381-385 (1957).

[34] Heinzer, W. Quotient overrings of an integral domain, Mathematika 17, 139-148 (1970).

[35] Heinzer, W., Ohm, J., and Pendleton, R. On integral domains of the form $\cap D_P$, P minimal, J. Reine Angew. Math. 241, 147-159 (1970).

[36] Jaffard, P. Dimension des anneaux de polynomes. La notion de dimension valuative, C. R. Acad. Sci. Paris Ser. A-B 246, 3305-3307 (1958).

[37] ______ Theorie de la Dimension dans les Anneaux de Polynomes, Gauthier-Villars, Paris, 1960.

[38] Jensen, C. U. On characterizations of Prüfer rings, Math. Scand. 13, 90-98 (1963).

[39] ________ Arithmetical rings, Acta Math. Acad. Sci. Hungar. 17, 115-123 (1966).

[40] Kaplansky, I. A characterization of Prüfer rings, J. Indian Math. Soc. (N.S.) 24, 279-281 (1960).

[41] Kirby, D. Components of ideals in a commutative ring, Ann. Mat. Pura Appl. (4) 71, 109-125 (1966).

[42] Krull, W. Beiträge zur Arithmetik kommutativer Integritätsbereiche, Math. Z. 41, 545-577 (1936).

[43] ________ Beiträge zur Arithmetik kommutativer Integritätsbereiche. II. v-Ideale und vollständig ganz abegeschlossene Integritätsbereiche, Math. Z. 41, 665-679 (1936).

[44] ________ Beiträge zur Arithmetik kommutativer Integritätsbereiche. VIII. Multiplikativ abgeschlossene Systeme von endlichen Idealen, Math. Z. 48, 533-552 (1943).

[45] Larsen, M. D. Equivalent conditions for a ring to be a P-ring and a note on flat overrings, Duke Math. J. 34, 273-280 (1967).

[46] Larsen, M. D., and McCarthy, P. J. Multiplicative Theory of Ideals, Academic Press, New York, 1971.

[47] Ohm, J. Integral closure and $(x, y)^n = (x^n, y^n)$, Monatsh. Math. 71, 32-39 (1967).

[48] Pendleton, R. L. A characterization of Q-domains, Bull. Amer. Math. Soc. 72, 499-500 (1966).

[49] Prüfer, H. Untersuchungen über die Teilbarkeitseigenschaften in Körpern, J. Reine Angew. Math. 168, 1-36 (1932).

[50] Richman, F. Generalized quotient rings, Proc. Amer. Math. Soc. 16, 794-799 (1965).

[51] Seidenberg, A. A note on the dimension theory of rings, Pacific J. Math. 3, 505-512 (1953).

[52] ________ On the dimension theory of rings II. Pacific J. Math. 4, 603-614 (1954).

[53] Storrer, H. H. A characterization of Prüfer rings, Canad. Math. Bull. 12, 809-812 (1969).

[54] Wadsworth, A. Noetherian pairs and the function field of a quadratic form., Thesis, Chicago, 1972.

INTEGRALLY CLOSED PAIRS

Edward D. Davis
Department of Mathematics
State University of New York at Albany
Albany, N. Y. 12222 U.S.A.

In the discussion following Gilmer's address Kaplansky introduced the notion "integrally closed pair", raised the question of characterizing these pairs and conjectured an approximate answer. Mott wisely conjectured that the methods of Davis' thesis [1] might well provide an answer. This note establishes the validity of both conjectures, at least in the Noetherian case, and somewhat more generally, for Krull domains.

INTEGRALLY CLOSED PAIRS OF KRULL DOMAINS. The reader will find an excellent discussion of "Krull domains" in Gilmer's book [2]; should he regard the notion as so much esoterica, he may substitute "integrally closed Noetherian domains". The pair of domains (A,B) is integrally closed if A is a subring of B and all intermediate rings (including A and B) are integrally closed. A bit of notation: for P a set of prime ideals of A, A_P means the localization of A with respect to P -- i.e., the intersection of the localizations of A at the primes of P. The parenthetical observation concluding the statement of the following theorem is this writer's understanding of Kaplansky's conjecture.

THEOREM 1. For A a Krull domain, the pair (A,B) is integrally closed if, and only if, one of the following holds: (1) A a field with B algebraic over A; (2) B a localization of A with respect to a set of height 1 primes that excludes only maximal such. The excluded set in (2) is unique, consisting of exactly those maximal M for which MB = B. (Consequently: For A Noetherian but not a field, if the pair (A,B) is integrally closed, then B is a localization of A with respect to a set of maximal ideals which is complementary to a uniquely determined set of invertible maximal ideals.)

The proof progresses via a sequence of lemmas for which we now make a blanket assumption: (A,B) is integrally closed. Since the polynomial algebra over a domain abounds in subalgebras which are not integrally closed, B is necessarily algebraic over A. This remark disposes of case (1); hereafter we assume A not a field. For S a multiplicative system in A, every ring between A_S and B_S is of the form C_S for C a ring between A and B; therefore the pair (A_S, B_S) is also integrally closed. We shall use this fact freely and without explicit comment.

LEMMA 1. B is contained in the quotient field of A.

Proof. We lose nothing by taking A local. Since B is algebraic over A, each element of B is of the form x/y with y in A and x integral over A; we show that x lies in A. For $z \neq 0$ a nonunit of A, observe that because A[zx] is integrally closed, A[x] = A[zx]. Consequently A[x] = A+zA[x]; whence A[x] = A by Nakayama's Lemma.

LEMMA 2. If A is local and x in B-A, then 1/x is in A. Thus: B is a ring of quotients of A in the local case, and in any case $B = A_{\mathcal{N}}$ for $\mathcal{N}$ the set of primes N for which $NB \neq B$.

Proof. That 1/x is in A follows from the observation that x lies in $A[x^2]$ -- consult the proof of theorem 1 of [1] for further details. Henceforth we assume A a Krull domain and use without explicit comment the following points: every localization of A is a Krull domain; each nonunit of A lies in at least one height 1 prime; $\mathcal{P}$ is the set of A's height 1 primes; $A = A_{\mathcal{P}}$, and for $\mathcal{M}$ and $\mathcal{N}$ subsets of $\mathcal{P}$, $A_{\mathcal{M}}$ contains $A_{\mathcal{N}}$ if, and only if, $\mathcal{N}$ contains $\mathcal{M}$; A_P is a valuation ring for each P in $\mathcal{P}$. The existence and unicity of a subset $\mathcal{M}$ of $\mathcal{P}$ such that $B = A_{\mathcal{M}}$ follow from these points and lemma 2; that $\mathcal{M}$ excludes only maximal ideals is a consequence of:

LEMMA 3. If A is local and $B \neq A$, then A is of dimension 1.

Proof. Since $B \neq A$, we lose nothing by assuming $B = A_M$ for M a subset of P excluding a single prime P. Then that B is a ring of quotients of A implies that $B = A_x$ for x in P but in no member of M. Were there a nonunit y not in P, then y/x would be in B-A, but x/y not in A -- a contradiction of lemma 2. Finally the sufficiency of (2):

LEMMA 4. For N the set of nonmaximal ideals belonging to P, the pair (A, A_N) is integrally closed.

Proof. Let M be a maximal ideal of an intermediate ring C. If $M \cap A$ is not in P, then $A_{M \cap A} = A_M$ for M the members of N contained in $M \cap A$; if $M \cap A$ is in P, then $A_{M \cap A}$ is a valuation ring. In either case $C_M = A_{M \cap A}$.

GENERALIZATIONS. Careful examination of the above arguments -- most importantly the proof of lemma 2 -- reveals more general results; we present selecta here, leaving the detailed exposition for publication elsewhere. Call a pair of domains (A,B) relatively integrally closed if A is a subring of B with all intermediate rings integrally closed in B. Except for the trivial case of A a field, an integrally closed pair (A,B) is merely a relatively integrally closed pair with B integrally closed. The central point of the generalization is that lemma 2 is valid in the more general setting; whence in the local case, relatively integrally closed pairs are compounded out of the two obvious examples: A a valuation ring, and the trivial example of A = B.

THEOREM 2. For A local, the pair (A,B) is relatively integrally closed if, and only if, $B = A_P$ for P a prime with $P = PA_P$ and A/P a valuation ring; for A not a field, the pair is integrally closed if, and only if, in addition A is integrally closed.

We now play a bit loose with two notions of Krull [3]: Call a domain fast-Noethersche if each of its nonzero nonunits admits a primary decomposition and also lies in at least one and at most a finite number of height 1 primes; call such a domain einbettungsfrei if its only primes associated in the sense of the primary decomposition to principal ideals are of height 1. Since a Krull domain is einbettungsfrei, theorem 1 is a special case of:

THEOREM 3. For fast-Noethersche A, the pair (A,B) is relatively integrally closed if, and only if, B is a localization of A with respect to a set of associated primes of principal ideals that excludes only such maximal ideals M for which A_M is a valuation ring (of dimension 1). The set of maximal ideals M for which MB = B can serve as the excluded set, which is unique for A einbettungsfrei. (Therefore: (A,B) is relatively integrally closed if, and only if, B is a localization of A with respect to a set of maximal ideals complementary to a set of maximal ideals M for which A_M is a valuation ring; this latter set is uniquely determined for A einbettungsfrei.)

REFERENCES

[1] Davis, E. D. Overrings of commutative rings. II. Integrally closed overrings, Trans. A.M.S. 110 (1964) 196-212.

[2] Gilmer, R. W. Multiplicative Ideal Theory, Queen's Papers on Pure and Applied Mathematics -- No. 12, Queen's University, Kingston, Ontario (1968).

[3] Krull, W. Einbettungsfreie, fast-Noethersche Ringe und ihre Oberringe, Math. Nachr. 21 (1960) 319-338.

NOETHERIAN INTERSECTIONS OF INTEGRAL DOMAINS II

William Heinzer*
Purdue University

If R and V are integral domains with quotient field K, we would like to consider the following questions: (i) when can one say that $D = R \cap V$ has quotient field K? (ii) when can one conclude that R or V is a localization of $D = R \cap V$? (iii) if R and V are noetherian, when can one say that $D = R \cap V$ is also noetherian?

A sample of what we can say on these questions is the following:

1) If R has nonzero Jacobson radical and V is a rank one valuation ring such that $D = R \cap V$ is irredundant, then both R and V are localizations of D.

2) With the same hypothesis as in (1), D is noetherian if and only if R and V are noetherian.

Result 1 is due to Bruce Prekowitz for the case when R is quasi-local, and was extended to the present form jointly by Jack Ohm and myself. Access to Prekowitz's manuscript was useful to us in our extension of the result. I am also indebted to Jack Ohm in that many of the other results given here were worked out jointly with Ohm.

Section 1 contains our results on questions (i) and (ii); while in Section 2 we deal with (iii), and also discuss (but do not resolve) the question of whether for R a 2-dimensional noetherian domain, must $R^* = \cap\{R_P | P$ is a height one prime of $R\}$ again be noetherian.

1. Some general remarks on intersections of integral domains. Throughout this section R and V denote integral domains with quotient field K. The intersection of the maximal ideals of R is called the <u>Jacobson radical</u> of R.

---

*This research was supported by NSF Grant GP-29326A1

<u>1.1 Lemma.</u> If V is a valuation ring on K and R has nonzero Jacobson radical J, then $D = R \cap V$ has quotient field K.

<u>Proof:</u> It will suffice to show that J is contained in the quotient field of D. If $a \in J$ and $a \in V$, then $a \in D$. If $a \in J$ and $a \notin V$, then $1 + a$ is a unit in R which is not in V, so $1/(1+a) \in R \cap V = D$.

<u>1.2 Remark.</u> If R has Jacobson radical zero, even for V a rank one valuation ring it need not follow that $D = R \cap V$ has quotient field K. For example, if k is a field, X an indeterminate, $R = k[X]$, $V = k[1/X]_{(1/X)}$, then $D = R \cap V = k$.

<u>1.3 Remark.</u> If $\{V_\alpha\}$ is a collection of subrings of some ring and if V_α has Jacobson radical J_α, then $T = \bigcap_\alpha V_\alpha$ has the property that $\bigcap_\alpha J_\alpha$ is contained in the Jacobson radical of T. This is clear from the characterization of the Jacobson radical of a ring A as $\{a \in A \mid 1 + ra$ is a unit in A for all $r \in A\}$, and the fact that any element of T which is a unit in each V_α is a unit in T.

<u>1.4 Remark.</u> It follows from 1.3 that if R and V both have nonzero Jacobson radical and $D = R \cap V$ again has quotient field K, then D has nonzero Jacobson radical. For D having quotient field K implies that nonzero ideals in R or V contract to nonzero ideals in D.

This yields the following extension of 1.1.

<u>1.5 Proposition.</u> If $\{V_i\}_{i=1}^n$ is a finite set of valuation rings on K and if R has nonzero Jacobson radical, then $D = R \cap (\bigcap_{i=1}^n V_i)$ has quotient field K.

In connection with 1.1 and 1.5, we would like to raise the following:

<u>1.6 Question.</u> If R and V are 1-dimensional quasi-local domains with quotient field K, must $D = R \cap V$ have quotient field K?

The following simple example shows the necessity of the 1-dimensionality assumption in 1.6.

1.7 Example. Let k be a field and let X and Y be indeterminates. Let $R = k[X, X^2Y]_{(X,X^2Y)}$ and $V = k[XY^2, Y]_{(XY^2,Y)}$. R and V are both 2-dimensional regular local rings having quotient field $k(X, Y) = K$, and are subrings of the formal power series ring $k[[X, Y]]$. Any element of R regarded as a formal power series in $k[[X, Y]]$ has the property that the X-degree of each monomial is greater than the Y-degree, and for any element of V the Y-degree is greater than the X-degree. Thus $R \cap V = k$.

Given that R has nonzero Jacobson radical and looking for conditions on V in order that $D = R \cap V$ has quotient field K, a natural hypothesis to put on V seems to be that V is a G-domain (an integral domain V is said to be a G-domain if the nonzero prime ideals of V have nonzero intersection, or equivalently if the quotient field of V is a finitely generated ring extension [7, p. 12].) I am aware of no example where R has nonzero Jacobson radical and V is a G-domain, but $D = R \cap V$ has smaller quotient field. It would follow from 1.8 that in such an example V may be assumed to be an intersection of rank one valuation rings.

We recall that an element α of the quotient field of V is said to be almost integral over V if $V[\alpha]$ is contained in a finite V-module or equivalently if the conductor of V in $V[\alpha]$ is nonzero.

1.8 Proposition. If V is a G-domain and $V \subset V^* \subset K$ with every element of V^* almost integral over V, then $D = R \cap V$ has quotient field K if (and obviously only if) $D^* = R \cap V^*$ has quotient field K.

Proof. Suppose D^* has quotient field K and let N be the contraction to D^* of the intersection of the nonzero primes of V^*. Since $V \subset V^* \subset K$, V^* is a G-domain, so N is a nonzero ideal of D^*. To see that D has quotient field K, it will suffice to show that N

is contained in the quotient field of D. If $\alpha \in N$, then some power of α, say α^n, is contained in the conductor of V in $V[\alpha]$. Thus α^n and α^{n+1} are in $R \cap V = D$, so $\alpha = \alpha^{n+1}/\alpha^n$ is in the quotient field of D.

1.9 Remark. If V is an integrally closed G-domain, then it is easily seen that the set of elements of K almost integral over V is precisely the intersection of the rank one valuation rings containing V [4, p. 359]. Hence, by repeated application of 1.8, if there exist G-domains R and V such that $R \cap V$ does not have quotient field K, then such R and V exist which are intersections of rank one valuation rings. Interesting examples of 1-dimensional quasi-local domains which are intersections of rank one valuation rings have been constructed by Nagata in [10] and Ribenboim in [14]. Perhaps these examples would be useful in answering 1.6.

We note the following obvious extension of 1.8.

1.10 Corollary. Let $V_1, .., V_n$ be G-domains with quotient field K and suppose $V_i \subset V_i^* \subset K$ with V_i^* almost integral over V_i. Then $D = R \cap V_1 \cap \ldots \cap V_n$ has quotient field K if (and obviously only if) $D^* = R \cap V_1^* \cap \ldots \cap V_n^*$ has quotient field K.

1.11 Corollary. Let $V_1, \ldots, V_n$ be G-domains with quotient field K such that each V_i is contained in only a finite number of rank one valuation rings of K. If R has nonzero Jacobson radical, then $D = R \cap V_1 \cap \ldots \cap V_n$ has quotient field K. In particular, $V_1 \cap \ldots \cap V_n$ has quotient field K.

Proof. Apply 1.5 and 1.10.

1.12 Remark. A 1-dimensional domain with only a finite number of prime ideals is a G-domain, so for example, it follows from 1.11 that if $V_1, \ldots, V_n$ are 1-dimensional semi-local (noetherian) domains with quotient field K, then $V_1 \cap \ldots \cap V_n$ has quotient field K and in fact $R \cap V_1 \cap \ldots \cap V_n$ has quotient field K whenever R has nonzero Jacobson radical.

We turn now to the question of when R or V is a localization of $D = R \cap V$.

<u>1.13 Proposition.</u> If R has nonzero Jacobson radical J and V is a valuation ring such that $J \not\subset V$, then R is a localization of $D = R \cap V$ and V is centered on a maximal ideal of D.

<u>Proof.</u> Let P denote the center of V on D. If $x \in J \setminus V$, then $1 + x$ is a unit of R, $1/(1+x) \in P$ and $x/(1+x) \in (D \cap J) \setminus P$. If $y \in R \setminus D$, then $y \notin V$, and if $x \in J \setminus V$, then $1/(1 + xy)$ and $y/(1 + xy)$ are in D and $1/(1 + xy)$ is a unit in R. Hence R is a localization of D. It remains to show that P is maximal. If $z \in D \setminus P$, $y \in (D \cap J) \setminus P$, and $a = 1/(1 + x)$ for $x \in J \setminus V$, then $a + yz \in P + yD$ is a unit in both R and V and hence a unit in D. Thus P is a maximal ideal of D.

<u>1.14 Remark.</u> If R has nonzero Jacobson radical J and V is a rank one valuation ring such that $R \not\subset V$, then $J \not\subset V$. For if $x \in R \setminus V$ and $y \in J$, $y \neq 0$, then $x^n y \notin V$ for some positive integer n.

<u>1.15 Proposition.</u> If R has nonzero Jacobson radical J, V is 1-dimensional quasi-local, $D = R \cap V$ is an irredundant intersection, and R is a localization of D, then V is centered on a maximal ideal P of D and $D_P = V$.

<u>Proof.</u> By assumption $R = D_S$ for some multiplicative system S in D. Irredundance of the intersection $D = R \cap V$ implies $S \cap P$ is non-empty. Let $x \in S \cap P$. If $y \in V$, then $y = a/b$ where $a, b \in J \cap D$, $b \neq 0$. Since V is 1-dimensional quasi-local, $x^n V < bV$ for n a sufficiently large positive integer and $b/(x^n + b)$ is a unit in V. Moreover, x^n a unit in R and $b \in J$ imply $x^n + b$ is a unit in R. Hence $\alpha = a/(x^n + b)$ and $\beta = b/(x^n + b)$ are in D and $\beta \notin P$. Since $y = \alpha/\beta$, $y \in D_P$, so $D_P = V$. It remains to show that P is

maximal. We note that $\beta = b/(x^n + b)$ is in $(J \cap D)\backslash P$. If $z \in D\backslash P$, then $\beta z + x$ is a unit in R and V, and hence a unit in D. Thus $P + zD = D$, so P is a maximal ideal in D.

We note the following consequence of 1.13, 1.14 and 1.15.

1.16 Corollary. If R has nonzero Jacobson radical and $V_1, \ldots, V_n$ are rank one valuation rings on K, then R is a localization of $D = R \cap V_1 \cap \ldots \cap V_n$. Moreover, irredundant V_i are centered on maximal ideals of D and are localizations of D.

We would like next to examine possibilities for generalization of 1.16 by weakening the hypothesis that the V_i are rank one valuation rings. The following two examples indicate some bounds on such generalization and help to justify the hypotheses of Proposition 1.19.

1.17 Example where R and V are 1-dimensional local domains and are not localizations of $D = R \cap V$ [2, p. 282]. Let k be a field of characteristic zero and let X, Y be indeterminates. Let $T = k[X, Y]_{(Y)} = k(X) + M$, where M is the maximal ideal of T. Let $R = k(X^2) + M$ and $V = k(X^2 + X) + M$. Then $D = R \cap V = k + M$, and R and V are not localizations of $k + M$. One can also in this way easily construct 1-dimensional quasi-local integrally closed domains R and V which are not localizations of $R \cap V$. Again begin with a rank one valuation ring of the form $F + M$ where F is a field, say $F = k(X_1, X_2)$ with X_1 and X_2 algebraically independent over k. Let $R = k(X_1) + M$ and $V = k(X_2) + M$, then R and V are 1-dimensional quasi-local integrally closed domains and are not localizations of $R \cap V = k + M$. Note that by 1.8 a construction of this form always yields a $D = R \cap V$ having the same quotient field. Also the R and V are dominated by a common valuation ring of their quotient field.

1.18 Example to show that in 1.13, if $J \subset V$, then R need not be a localization of D. Let k be a field and let X, Y be indeterminates. Let $T = k[X, Y]_{(Y)} = k(X) + M$, where M is the maximal ideal of T.

Let $R = k[X] + M$ and $V = k[1/X]_{(1/X)} + M$. Then M is the Jacobson radical of R, V is a valuation ring, and $D = R \cap V = k + M$, so R is not a localization of D.

<u>1.19 Proposition.</u> If R has nonzero Jacobson radical J, and V is a G-domain such that V is contained in only a finite number of rank one valuation rings of K, say $V_1, \ldots, V_n$, and if moreover $R \not\subset V_i$ for each i, then R is a localization of $D = R \cap V$.

<u>Proof:</u> By 1.16, R is a localization of $D' = R \cap V_1 \cap \ldots \cap V_n$, say $R = D'_S$. Since $R \not\subset V_i$, there exists $s_i \in S$ such that s_i is in the maximal ideal of V_i. Let $s = s_1 \ldots s_n$. Then s is in the maximal ideal of each V_i. Since V is a G-domain, every nontrivial valuation ring of K containing V is contained in a rank one valuation ring of V. It follows that s is in the maximal ideal of every nontrivial valuation ring of K containing V. Therefore s is integral over V and some power of s, say s^n is contained in the conductor of V in $V[s]$. Thus $s^n \in V \cap R$ and $1/s^n \in R$. Hence $D[1/s^n] = R \cap V[1/s^n] = R$.

<u>1.20 Corollary.</u> If R has nonzero Jacobson radical and V is 1-dimensional quasi-local such that V is contained in only a finite number $V_1, \ldots, V_n$, of rank one valuation rings of K, then $R \not\subset V_i$ for each i implies R and V are both localizations of D and V is centered on a maximal ideal of D.

<u>Proof:</u> Apply 1.19 and 1.15.

Without the assumption that R has nonzero Jacobson radical, it can happen for V a rank one valuation ring not having rational value group that $R \cap V = D$ is irredundant, D again has quotient field K, but V is not a localization of D [12, p. 330]. We can give, however, the following corollary to 1.13 and 1.14.

1.21 Corollary. If V is a rank one valuation ring, $R \cap V = D$ is an irredundant intersection, D again has quotient field K, and V is centered on a height one prime P of D, then $D_P = V$.

Proof. If $S = D \setminus P$, then $D_P = R_S \cap V_S = R_S \cap V$. Thus irredundance of $R \cap V$ implies that $D_P = V$ is equivalent to $R_S = K$; and $R_S \neq K$ implies irredundance of $R_S \cap V = D_P$. Since P is a height one prime, D_P is a G-domain, so R_S is also a G-domain. Hence, if $R_S \neq K$, then R_S would have nonzero Jacobson radical which by 1.13 and 1.14 would imply V is a localization of D_P. It follows that $R_S = K$ and $D_P = V$.

2. Noetherian corollaries and a question on noetherian intersections. We can now quickly give some results in the spirit of [5] on noetherian properties of $D = R \cap V$. We continue to use R and V to denote integral domains with quotient field K.

2.1 Theorem. If R has nonzero Jacobson radical and $V_1, \ldots, V_n$ are rank one valuation rings with quotient field K, then $D = R \cap V_1 \cap \ldots \cap V_n$ is noetherian if and only if R and the irredundant V_i are noetherian.

Proof. By 1.16, R and the irredundant V_i are localizations of D, so D noetherian implies R and the irredundant V_i are noetherian. Also by 1.16, each irredundant V_i is centered on a maximal ideal of D. Thus [5, Theorem 1.10] yields that if R and the irredundant V_i are noetherian, then D is noetherian.

2.2 Theorem. If R has nonzero Jacobson radical and V is 1-dimensional quasi-local such that V is contained in only a finite number $V_1, \ldots, V_n$ of rank one valuation rings of K, and if moreover $R \not\subset V_i$ for each i, then $D = R \cap V$ is noetherian if and only if R and V are noetherian.

Proof: By 1.20, R and V are both localizations of D, so D noetherian implies R and V are noetherian. Also by 1.20, V is centered on a maximal ideal P of D. Since for any multiplicative system S of D, $D_S = R_S \cap V_S$ (see for example [5, Lemma 1.1]), if Q is a prime of D such that $Q \neq P$, then $R \subset D_Q$. It now follows easily that if R and V are noetherian, then D is noetherian; for example, apply [5, Corollary 1.8].

2.3 Remark. Concerning possible generalizations of Theorem 2.2, we note first the following. A simple example such as [5, Example 1.19] shows that one can have $D = R \cap V$ with R and D being 2-dimensional local (noetherian) domains and V being 1-dimensional quasi-local but not noetherian. Also it is easy to construct an example of the form $D = R \cap V$ where D and R are semi-local (noetherian) and V is a valuation ring of rank > 1. Finally, Example 1.17 shows that without the hypothesis that R and V have no common overring < K, it can happen that R and V are 1-dimensional local (noetherian) and $D = R \cap V$ is not noetherian.

We can give the following corollary to Theorem 2.2.

2.4 Corollary. If R is noetherian with nonzero Jacobson radical and V is 1-dimensional semi-local (noetherian) such that R and V have no common overring < K, then $D = R \cap V$ is a noetherian domain.

Proof. If $P_1, \ldots, P_n$ are the maximal ideals of V, let $V_i = V_{P_i}$. Then $D = R \cap V_1 \cap \ldots \cap V_n$. By 2.2 and 1.20, $R' = R \cap V_1$ is noetherian, V_1 is centered on a maximal ideal P of R' and if Q is a prime ideal of R' distinct from P, then $R \subset R'_Q$. Hence if W is any quasi-local domain such that $R' \subset W$, then either $R \subset W$ or $V_1 \subset W$. Therefore our hypotheses insure that R' and V_2 have no common overring < K, and a simple induction argument yields that $D = R \cap V_1 \cap \ldots \cap V_n$ is noetherian.

<u>2.5 Remark.</u> The simplest examples of integral domains with nonzero Jacobson radical are, of course, those with only a finite number of maximal ideals. However, in general, if $\mathfrak{A}$ is a nonzero ideal in R, then $S = \{1 + a | a \in \mathfrak{A}\}$ is a multiplicative subset of R and $\mathfrak{A}R_S$ is contained in the Jacobson radical of R_S, so the localization R_S has nonzero Jacobson radical. Since no prime ideal of R containing $\mathfrak{A}$ is lost in R_S, this sort of localization easily yields noetherian domains with nonzero Jacobson radical for which the maximal spectrum has large dimension. Also as Graham Evans pointed out to me, a very simple way to obtain domains with nonzero Jacobson radical is by taking formal power series rings. Thus the maximal spectrum of any noetherian domain T can be realized as the maximal spectrum of a noetherian domain R with nonzero Jacobson radical by setting $R = T[[X]]$.

We should like to conclude with some comments on the following question, cf. Krull [8].

<u>2.6 Question.</u> If R is a 2-dimensional noetherian domain and $R^* = \cap\{R_P | P$ is a height one prime of $R\}$, then must R^* again be noetherian?

Ferrand and Raynaud prove in [3, Corollary 1.4, p. 298] that if R is a 2-dimensional local (noetherian) domain and $R^* = \cap \{R_P | ht(P) = 1\}$ is integral over R, then R^* is noetherian. It follows readily by a simple localization argument that if R is a 2-dimensional semi-local domain and R^* is integral over R, then R^* is noetherian. In fact, the assumption that R^* is integral over R may be deleted. This can be seen as follows: let A be the integral closure of R. A is then a 2-dimensional semi-local domain [9, p. 118 and p. 120]. If all maximal ideals of A have height 2, then $R^* \subset A$, so R^* is noetherian. If A has some maximal ideals of height 1, then choose $x \in A$ such that the maximal ideals of A that contain x are precisely the maximal ideals of height 1. It follows from [6, Theorem 2.7, p. 150] that $A[1/x]$ is the integral closure of R^*. Let B denote a finitely generated ring extension of R obtained by adjoining to R the coefficients of an equation of integral dependence for $1/x$ over R^*. Then B is a semi-local domain with integral closure $A[1/x]$, and $B^* = \cap\{B_P | P$ is a height one prime of $B\} = R^*$. Since B^* is integral over B, the Ferrand-Raynaud result implies that $B^* = R^*$ is noetherian.

Thus for R a 2-dimensional noetherian domain, it is at least true that R^* localized at any prime ideal is noetherian. The following proposition will show that R^* has noetherian spectrum and that R^* must be noetherian when all domains between R and the integral closure of R are noetherian.

2.7 Proposition. Let R be a 2-dimensional noetherian domain, let $R^* = \cap\{R_P | P$ is a height one prime of $R\}$, and let A denote the integral closure of R. If $R^* \not\subset A$, and if $\{P_i\}$ denotes the set of maximal ideals of A which are also of height one, then R^* is noetherian if and only if $T = R^* \cap \{A_{P_i}\}$ is noetherian. Moreover, $T \subset A$, and R^* is a flat T-module. In particular, R^* has noetherian spectrum.

Proof. If M is a maximal ideal of A of height 2, then $A_M = \cap\{A_Q | Q$ is a height one prime of A contained in $M\}$, and $Q \cap R = P$ is a height one prime of R such that $R_P \subset A_Q$. Thus $R^* \subset A_M$. It follows that $T \subset A$. If P_i is a height one maximal ideal of A, then $P_i \cap T$ is also maximal, so the noetherian valuation ring A_{P_i} is centered on a maximal ideal of T. It therefore follows from [5, Theorem 1.10] that R^* is a flat T-module, and that R^* is noetherian if and only if T is noetherian. Since A is a noetherian domain and is an integral extension of T, T has noetherian spectrum. Finally, since R^* is a flat extension of T contained in the quotient field of T, T having noetherian spectrum implies R^* has noetherian spectrum.

We close with the following example which shows that between a 2-dimensional noetherian domain R and the integral closure of R there can exist a non-noetherian domain B such that B localized at any prime ideal is noetherian.

2.8 Example. Let T be a 1-dimensional noetherian domain such that the integral closure of T is not a finite T-module, but such that for each prime ideal Q of T, the integral closure of T_Q is a finite T_Q-module. (Such a domain T may be constructed, for example, as in Nagata [11, p. 35] or Dade [1].) Let X be an indeterminate and let R = T[X]. Then, as Kaplansky pointed out to me, it is easy to see the existence of a non-noetherian domain B between R and the integral closure of R: let $\{T_i\}_{i=1}^{\infty}$ be a strictly ascending chain of domains between T and the integral closure of T, and let $B = R[\{T_i X^i\}_{i=1}^{\infty}]$. Then $(T_1X) < (T_1X, T_2X^2) < \ldots < (T_1X, \ldots, T_nX^n) < \ldots$ is a strictly ascending chain of ideals of B, so B is not noetherian. However, for any prime ideal P of B, if $Q = P \cap T$, then B_P is a localization of an integral extension of $T_Q[X]$. Since the integral closure of $T_Q[X]$ is a finite module, B_P must be noetherian.

REFERENCES

1. E. C. Dade, Rings in which no fixed power of ideal classes becomes invertible, Math Annalen 148(1962), 65-66.

2. P. Eakin, The converse to a well known theorem on noetherian rings, Math Annalen 177(1968), 278-282.

3. D. Ferrand and M. Raynaud, Fibres formelles d'un anneau local noethérien, Ann. scient. Ec. Norm. Sup. 3 (1970), 295-311.

4. R. Gilmer and W. Heinzer, On the complete integral closure of an integral domain, J. Australian Math. Soc. 6(1966), 351-361.

5. W. Heinzer and J. Ohm, Noetherian intersections of integral domains, Trans. Amer. Math. Soc. 167(1972).

6. W. Heinzer, J. Ohm, and R. Pendleton, On integral domains of the form $\cap D_P$, P minimal, J. Reine Angew. Math. 241(1970), 147-159.

7. I. Kaplansky, Commutative Rings, Allyn and Bacon, Boston (1970).

8. W. Krull, Einbettungsfreie fast-Noetherche Ringe und ihre Oberringe, Math. Nachr. 21(1960), 319-338.

9. M. Nagata, Local Rings, Interscience, New York (1962).

10. M. Nagata, On Krull's conjecture concerning valuation rings, Nagoya Math. J. 4(1952), 29-33.

11. M. Nagata, On the closedness of singular loci, Publs. Math. Inst. Hautes Etudes Sci. 2(1959), 29-36.

12. J. Ohm, Some counterexamples related to integral closure in D[[X]], Trans. Amer. Math. Soc. 122(1966), 321-333.

13. B. Prekowitz, Intersections of quasi-local domains, Chicago Ph.D. thesis, 1971.

14. P. Ribenboim, Sur une note de Nagata relative á un problème de Krull, Math. Zeit. 64 (1956), 159-168.

COHEN-MACAULAY MODULES

Melvin Hochster[1]
University of Minnesota
Minneapolis, Minnesota 55455

The object of this paper is to discuss the conjecture, which will be abbreviated (E), that every complete local ring of dimension n possesses a finitely generated module of depth n. It is noted that several conjectures which have been open for some time follow from (E), and the connection of (E) with Serre's conjecture on multiplicities over regular local rings is discussed. In fact Serre's conjecture is proved for dimension ≤ 4 using the ideas under consideration.

A number of proofs of (E) for the two-dimensional case are given, and some possible methods for handling the general case are discussed. One of these is proposed as particularly worthy of study and is applied to an interesting class of examples in dimension 3 to obtain modules of depth 3. These examples do not yield easily to other techniques.

1. THE BASIC FACTS AND THE BASIC CONJECTURE

Throughout, all rings are assumed to be commutative, with identity, ring homomorphisms are assumed to preserve the identity, and modules are assumed to be unital. By a <u>local ring</u> we mean a Noetherian ring with a unique maximal ideal. If M is an R-module, we say that $r_1, \ldots, r_k$ in R form an M-<u>sequence</u> or a <u>regular sequence</u> on M if $(r_1, \ldots, r_k)M \neq M$ and r_i is not a zerodivisor on $M/(r_1, \ldots, r_{i-1})M$, $1 \leq i \leq k$. If I is an ideal of R and M is an R-module, depth (I,M) or $D(I,M)$ denotes the supremum of lengths of M-sequences contained in I, provided $IM \neq M$. If $IM = M$ (e.g. if $I = R$ or $M = 0$) we make the convention that $D(I,M) = \infty$. If (R,P) is local and M is finitely generated, then $D(M) = D(P,M)$ is finite if $M \neq 0$. If I is an ideal of R we write $G(I)$ or grade I for $D(I,R)$. If R is Noetherian and $I \neq R$, then $G(I) < \infty$.

---

[1]This research was supported in part by National Science Foundation Grant GP-29224X.

Let M be a finitely generated module over a local ring (R,P). Then $\ell(M) < \infty$ if and only if $\mathrm{Ann}_R M$ is P-primary, where ℓ denotes length. More generally, if $\dim M =$ Krull $\dim(R/\mathrm{Ann}_R M)$, we have that the following conditions are equivalent:

1) k is the least possible integer such that $\ell(M/(r_1, \ldots, r_k)M) < \infty$.

2) The images of $r_1, \ldots, r_k$ form a system of parameters for $R/\mathrm{Ann}_R M$.

3) For each i, $1 \leq i \leq k$, $\dim M/(r_1, \ldots, r_k)M = \dim M - i$.

A sequence $r_1, \ldots, r_k$ satisfying these equivalent conditions is called a <u>system of parameters</u> for M, and then $k = \dim M$. See Serre [16], Ch. IIIA.

A finitely generated module M over a local ring (R,P) is called <u>Cohen-Macaulay</u>, or C-M, if $D(M) = \dim M$. The condition is equivalent to the statement that every system of parameters for M is an M-sequence. The condition is preserved by localization at a smaller prime in $\mathrm{Supp}\, M$. If $I \subseteq \mathrm{Ann}_R M$, M is C-M over R if and only if it is C-M over R/I. When R is not local, we say that a finitely generated module M is <u>Cohen-Macaulay</u> if M_P is C-M over R_P for every P in $\mathrm{Supp}\, M$.

If R is local and M is a finitely generated R-module, then $D(M) \leq \dim R$ (in fact, $D(M) \leq \dim M$). It is natural to ask whether the upper bound $\dim R$ for $D(M)$ can be achieved. If it is achieved for a certain M, we say that M is a <u>maximal</u> C-M module over R. In case R is a domain, a maximal C-M module is the same as a faithful C-M module. In general, a maximal C-M module is torsion-free.

If $\dim R \leq 1$, one can produce a maximal C-M module by dividing out by a prime of maximal coheight. In general, this trick reduces the problem to the domain case. But in general, even in dimension 2, the answer is negative.

To see this, suppose R possesses a C-M module M with $\mathrm{Supp}\, M = \mathrm{Spec}(R)$. Then every quotient of a polynomial or formal power series ring over R satisfies 1) the saturated chain condition, and 2) Grothendieck's (CMU) condition, [6] p.162, so that if S is the ring, $\{Q \in \mathrm{Spec}(S) : S_Q \text{ is C-M}\}$ is Zariski open in $\mathrm{Spec}(S)$. In fact, if $S = R[x_1, \ldots, x_n]$ or $R[[x_1, \ldots, x_n]]$, then $M \otimes_R S$ is C-M over S with support $\mathrm{Spec}(S)$, and facts 1) and 2) are immediate from

Serre [16], p. IV-24, Corollaire 3, and Grothendieck [6], p. 163, Remarque (6.11.9) (ii), respectively. The two-dimensional local domain R_0 discussed by Kaplansky in [11], §§ 4 and 23 (of course, the example is originally due to Nagata), fails to satisfy 1), while the two-dimensional local domain discussed by Ferrand and Raynaud [5], Proposition 3.5, fails to satisfy 2).

However, if R is complete both 1) and 2) are automatic because R is a quotient of a regular local ring. This author conjectures:

Conjecture (E). *If R is a complete local ring, R possesses a maximal Cohen-Macaulay module.*

Equivalently:

Conjecture (E'). *If R is a complete local domain, R possesses a faithful Cohen-Macaulay module.*

If S is a complete local domain, we know (see Nagata [13], p. 109, Corollary (31.6)) that S is a module-finite domain extension of a complete regular local ring R. If M is a finitely generated S-module, then M is also a finitely generated R-module, and $D_S(M) = D_R(M)$, either from Serre [16], p. IV-18, Proposition 12 or from Levin and Vasconcelos [12], p. 317, Lemma. Over R, every module has finite projective dimension and if pd denotes projective dimension, then for every finitely generated R-module E, $D_R(E)+\text{pd}_R E = D(R) = \dim R$. Hence, $D_S(M) = \dim S = \dim R$ if and only if $D_R(M) = D(R)$, i.e. if and only if $\text{pd}_R M = 0$. This says that M is an S-module which is a nonzero free R-module. It follows that an equivalent conjecture to (E) and (E') is:

Conjecture (E"). *If R is a complete regular local ring and S is a module-finite domain extension, then for some integer $n \geq 1$ R^n has the structure of an S-module.*

To give R^n an S-module structure (by which we mean, of course, an S-module structure which extends its given R-module structure) is equivalent to giving a ring homomorphism of S into the ring $\mathfrak{M}_n(R)$ of n by n matrices over R which extends the usual identification of R with the scalar matrices. This enables us to reformulate the conjecture thus:

Conjecture (E'''). *If R is a complete regular local ring and S is a module-finite extension domain, then for some integer $n \geq 1$, the usual ring homomorphism $R \to \mathfrak{M}_n(R)$ extends to a ring homomorphism $S \to \mathfrak{M}_n(R)$.*

A finitely generated module M over a Noetherian ring R is called *perfect* if $M \neq 0$ and $G(\mathrm{Ann}_R M) = \mathrm{pd}\, M$ (so that, in particular, $\mathrm{pd}\, M < \infty$). Recall that, in general, if $\mathrm{pd}\, M$ is finite and $M \neq 0$, then $G(\mathrm{Ann}_R M) \leq \mathrm{pd}\, M$; see Rees [15], p. 30. If R is a complete local domain, R can be represented as S/Q, where S is a complete regular local ring and Q is prime. A maximal C-M R-module is then precisely the same as a perfect S-module with annihilator Q. Hence, we can state one last form of the basic conjecture:

Conjecture (E''''). *If S is a complete regular local ring and Q is a prime ideal of S, then there exists a perfect S-module M such that $\mathrm{Ann}_S M = Q$.*

2. CONSEQUENCES OF CONJECTURE (E)

We now discuss a number of open questions which are settled affirmatively by (E). The first of these can be regarded as a homological version of Krull's principal ideal theorem.

Conjecture (A). *Let $R \to S$ be a homomorphism of Noetherian rings, let M be an R-module, and let $I = \mathrm{Ann}_R M$. Let Q be a minimal prime of IS. Then $\mathrm{ht}\, Q \leq \mathrm{pd}_R M$.*

We shall use ht Q for $\dim R_Q$, and for an arbitrary ideal J, we use ht J for $\inf\{\mathrm{ht}\, Q : J \subset Q,\ Q \text{ prime}\}$.

To understand conjecture (A) better, consider the example where $R = Z[x_1, \ldots, x_n]$, $I = (x_1, \ldots, x_n)$, and $M = R/I$. In this case, $\mathrm{pd}_R M = n$. To give a homomorphism $R \to S$ is the same as to specify n elements $s_1, \ldots, s_n$

of S to serve as the images of $x_1, \ldots, x_n$. Then $IS = (s_1, \ldots, s_n)$, and the conjecture asserts that the height of any minimal prime of an ideal generated by n elements is at most n.

On the other hand, if we take $R = S$ and note that grade $I \leq \text{ht } P$ for any minimal prime P of I, we can conclude another known result from (A): Rees' theorem [15], p. 30 that $G(\text{Ann}_R M) \leq \text{pd}_R M$ if $M \neq 0$. One interesting point here is that the generalization of Rees' theorem which allows the change of ring is true (Theorem 1 of [8]) even in a non-Noetherian version, and this permits us to conclude (E) $\Rightarrow$ (A) ; in fact, a very weak form of (E) suffices. We return to this point soon.

The so-called "intersection theorem" is an easy corollary of (A) : the converse is less trivial but also true. We first recall the intersection theorem:

Conjecture (A'). *If* (R,P) *is a local ring and* M, N *are finitely generated* R-*modules such that* $\ell(M \otimes N) < \infty$, *then* $\dim N \leq \text{pd } M$.

This result is proved in Peskine and Szpiro [14] in many interesting special cases: if R has characteristic $p > 0$, if R is a localization of a ring finitely generated over a field, if $\text{pd } M \leq 2$, and, of course, if R is regular the result was known, Serre [16], p. V-18, Théorème 3. It is trivial to reduce to the case where N is prime cyclic, i.e. $N \cong R/Q$, where Q is prime.

Proposition 2.1. *Conjecture* (A) *holds if and only if Conjecture* (A') *holds*.

Proof. Assume (A). It suffices to prove (A') when $N = R/Q$. Let $S = R/Q$. The hypothesis $\ell(M \otimes N) < \infty$ is equivalent to the hypothesis Ann M + Ann N is P-primary. Let I = Ann M. Then I+Q is P-primary, and IS is primary to the maximal ideal of S. Hence, $\dim N = \dim S = \text{ht } P/Q \leq \text{pd}_R M$, by (A).

Now assume (A'). Let $R \to S$, M, I and Q as in the statement of (A) be given. If P is the inverse image of Q, we can consider the local homomorphism $R_P \to S_Q$ instead of the original one, and we can complete. Thus, we can assume that $(R,P) \to (S,Q)$ is a local homomorphism of complete local rings and that IS is primary to Q, and we must show that $\text{ht } Q \leq \text{pd}_R M$. By making faithfully flat local extensions of R and S we can get a coefficient ring of R to map onto a coefficient ring of S , and by adjoining analytic indeterminates to R we can then further guarantee that the map $R \to S$ is onto, i.e. that $S = R/J$. We can then apply (A') with $N = R/J$.

By modifying this argument, we can deduce that (A) holds in certain situations where (A') is known to hold locally, e.g. if R has characteristic $p > 0$, if R , S are finitely generated k-algebras for some field k , if R is regular, or if $\mathrm{pd}_R M \leq 2$.

We next want to show that (E) does, in fact, imply (A) . We shall even show that the weaker conjecture (E*) below implies (A) .

Conjecture (E*). _If (R',P') is a complete local domain, there is a (possibly non-Noetherian) R-module T such that_ $P'T \neq T$ _and_ $D(T) = \dim R'$.

Proposition 2.2. _Conjecture (E*) implies Conjecture (A)_ .

Proof. It suffices to prove (A') when $N = R/Q$ is prime cyclic. Let T be as in Conjecture (E*) for $R' = R/Q$. Then by Theorem 1 of [8], $\dim N = D(P/Q, T) = D(IR', T) \leq \mathrm{pd}\ M$.

Let us say that a local ring R has _abundant_ C-M modules if for every prime Q of R there is a C-M module over R whose annihilator is Q .

Corollary 1.3. _Conjecture (A') holds for a local ring R if R has abundant Cohen-Macaulay modules._

Note the following corollary of Conjecture (A) :

Conjecture (A_1). _If R is a local ring which possesses a module M of finite length and finite projective dimension, then R is Cohen-Macaulay._

(A') $\Rightarrow$ (A_1) because we may take $N = R$ and then $\dim R \leq \mathrm{pd}\ M \leq D(R)$, as required.

R. Y. Sharp [17] has recently proved a formal duality between finitely generated modules of finite injective dimension and finitely generated modules of finite projective dimension over any C-M local ring R which is a quotient of a Gorenstein local ring, and, in particular, over any complete C-M local ring. On the other hand, Bass has conjectured [1]:

Conjecture (B). If a local ring (R,P) possesses a finitely generated module of finite injective dimension, then R is C-M.

This can be reduced to the complete case and, to some extent, is dual to Conjecture (A_1). Peskine and Szpiro [14] have proved (B) in roughly the same cases that they have proved (A') . They also note in [14] that a weak form of (E) implies (B) : one needs a module of maximal depth n over a local extension R' of R such that $\dim R' = \dim R (= n)$ and $\dim R'/PR' = 0$. (The module is to be finitely generated as an R'-module.)

We now want to discuss multiplicities. If (R,P) is local and M, N are finitely generated R-modules with $\operatorname{pd} M < \infty$, then if $\ell(M \otimes N) < \infty$ (i.e. $\operatorname{Ann} M + \operatorname{Ann} N$ is P-primary), we have that $\operatorname{Tor}_i(M,N)$ is annihilated by $\operatorname{Ann} M + \operatorname{Ann} N$ for each i, and each $\ell(\operatorname{Tor}_i(M,N)) < \infty$, so that

$$e_R(M,N) = e(M,N) = \Sigma_i \ (-1)^i \ell(\operatorname{Tor}_i(M,N))$$

is a well-defined integer, which we refer to as the intersection multiplicity of M, N. If R is regular, one knows that under these hypotheses $\dim M + \dim N \leq \dim R$, Serre [16], p. V-18 (and it is conjectured that this holds without the regularity: $\operatorname{pd} M < \infty$ and $\ell(M \otimes N) < \infty$ should be enough), and one also knows that if the completion of R is a formal power series ring over a discrete valuation ring (in particular, if R is unramified) then $e(M,N) \geq 0$ and $e(M,N) > 0$ if and only if $\dim M + \dim N = \dim R$. Serre has conjectured that this holds for all regular local rings [16], p. V-14. This question is still open. We shall prove below that if both (E) and an additional conjecture (F) hold[2],

[2] At the time of this talk, I mistakenly thought that I could prove that (E) implies Serre's conjecture without assuming (F).

then Serre's conjecture holds. We first state Serre's conjecture formally:

Conjecture (C). Let (R,P) be a regular local ring and let M, N be finitely generated R-modules such that $\ell(M\otimes N) < \infty$. Then $e(M,N) \geq 0$ and $e(M,N) > 0$ if and only if $\dim M + \dim N = \dim R$.

We next want to discuss Conjecture (F): afterwards we shall prove (E) + (F) $\Longrightarrow$ (C).

Let E be a perfect module over $A = Z[x_1, \ldots, x_n]$. If R is an A-algebra such that $G(IR) \geq G(I)$, where $I = \mathrm{Ann}_A E$, but $IR \neq R$, then $R\otimes_A E$ is a perfect R-module of the same projective dimension as E. In fact, if $\mathcal{K}$ is a projective resolution of E, then $R\otimes_A \mathcal{K}$ is a projective resolution of $E' = R\otimes_A E$ (i.e. $\mathrm{Tor}_i^A(R,E) = 0$, $i \geq 1$). See [8]. In this situation, we shall say that E' is g-perfect. The significance of this notion for the study of multiplicities is apparent from the following:

Lemma 2.4. Let (R,P) be a Cohen-Macaulay ring and let M be a g-perfect R-module. Let N be a finitely generated R-module such that $\ell(M\otimes N) < \infty$. Then:

a) $\dim M + \dim N \leq \dim R$,

b) $e(M,N) \geq 0$, and

c) $e(M,N) > 0$ if and only if $\dim M + \dim N = \dim R$.

Proof. As usual, we can assume that R is complete. Write $R = S/J$, where S is an unramified regular local ring. Suppose $M = R\otimes_A E$, where E is perfect over $A = Z[x_1, \ldots, x_n]$. The homomorphism $A \to R$ lifts to a homomorphism $A \to S$. Let $M^* = S\otimes_A E$. We shall show that M^* is also g-perfect. Let $I = \mathrm{Ann}_A E$. We need to show that $G(IS \geq G(IR)$. But $G(IR) = G((J+IS)/J) = \mathrm{ht}\,(J+IS)/J = \mathrm{ht}\,(J+IS) - \mathrm{ht}\,J \leq \mathrm{ht}\,J + \mathrm{ht}\,IS - \mathrm{ht}\,J = \mathrm{ht}\,IS = G(IS)$. Let $\mathcal{K}$ be a projective resolution of E over A, so that $M^*\otimes_A \mathcal{K}$ is a projective resolution of M^* over S, and $R\otimes_S(M^*\otimes_A \mathcal{K})$ is a projective resolution of $R\otimes_S M^* = M$ over R. Regard N as an S-module. Since $IN = 0$, $(R\otimes_S(M^*\otimes_A\mathcal{K}))\otimes_R N = (M^*\otimes_A\mathcal{K})\otimes_S N$. Thus, $\mathrm{Tor}_i^S(M^*,N) = \mathrm{Tor}_i^R(M,N)$, and $e_R(M,N) = e_S(M^*,N)$. Since Conjecture (C) holds for unramified regular local rings, all three facts will follow if $\dim R - \dim M - \dim N = \dim S - \dim M^* - \dim N$. But $\dim R - \dim M = D(R) - \operatorname{depth} M = \operatorname{pd} M = \operatorname{pd} M^* = D(S) - \operatorname{depth} M^* = \dim S - \dim M^*$, as required.

This motivates the following definitions. Let (R,P) be a local C-M ring. Let $\underline{G}(R)$ be the Grothendieck group of R-modules of finite length and finite homological dimension. Let $g(R)$ be the subgroup of $\underline{G}(R)$ generated by the classes $[E]$ of the g-perfect modules E of finite length. Let $\overline{G}(R) = \underline{G}(R)/g(R)$. Let $R_n = R(t_1, \ldots, t_n)$ $(= R[t_1, \ldots, t_n]_{PR[t_1, \ldots, t_n]})$. We have maps $\underline{G}(R_m) \to \underline{G}(R_n)$, $m \leq n$, by $[E] \to [E \otimes_{R_m} R_n]$. The maps take $g(R_m)$ into $g(R_n)$ for each m, n, and we therefore have maps $\overline{G}(R_m) \to \overline{G}(R_n)$, $m \leq n$. Let $F(R) = \varinjlim_n \overline{G}(R_n)$.

Conjecture (F). <u>For each Cohen-Macaulay local ring</u> R , $F(R) = 0$.

For applications to multiplicities, we only need this conjecture when R is a local complete intersection.

We shall prove (F) for the case $\dim R \leq 2$.

If $\dim R = 0$, every module of finite projective dimension is free and there is nothing to prove.

If $\dim R = 1$, and M is a perfect module with $\operatorname{pd} M = 1$, then $M = \operatorname{Coker}(R^n \xrightarrow{f} R^n)$ for some n , where $\det f$ is not a zerodivisor. M itself is then g-perfect: let $A = Z[x_{ij}: 1 \leq i,j \leq n]$, and map A to R by taking the entries of (x_{ij}) to the corresponding entries of the matrix of f . Let $f^*: A^n \to A^n$ have matrix (x_{ij}) , and let $E = \operatorname{Coker} f^*$. Then M arises from E in the required manner.

To obtain the result for dimension 2 we use a result of Buchsbaum and Eisenbud [3] . Let (x_{ij}) be a b_1 by b_0 matrix of indeterminates over Z, $b_1 \geq b_0$. Let $A = Z[x_{ij}]_{ij}$. Let $f: A^{b_1} \to A^{b_0}$ be the map whose matrix is (x_{ij}) . Let $E(b_1, b_0) = \operatorname{Coker} f$.

Lemma 2.5 (Buchsbaum-Eisenbud). $E(b_1, b_0)$ is perfect of projective dimension b_1-b_0+1. Its annihilator is the ideal of maximal minors of (x_{ij}), and that ideal has grade b_1-b_0+1.

If $b_1-b_0 = 1$, the Betti numbers of E are $b_0, b_1, 1$, i.e. $b_0, b_0+1, 1$.

If $b_1-b_0 = 2$, the Betti numbers of E are b_0, b_0+2, b_0+2, b_0.

In fact, in [3] Buchsbaum and Eisenbud give explicit minimal resolutions of these modules, which they refer to as "generic torsion modules". It follows that if (x_{ij}) has entries in P, where (R,P) is local, and is a b_1 by b_0 matrix, $b_1 \geq b_0$, then if the grade of the ideal of maximal minors of (x_{ij}) is b_1-b_0+1 then $\mathrm{Coker}\,(x_{ij})$ is perfect of projective dimension b_1-b_0+1 and has the same Betti numbers (and, in fact, essentially the same resolution) as a matrix of indeterminates does in the generic case.

In the sequel we want to consider what we shall refer to as resolutions with matrices in "general position". Consider a free complex $\mathcal{K} = 0 \to R^{b_n} \to \dots \to R^{b_0} \to 0$ which is a resolution of M over R, i.e. the homology vanishes except for $H_0(\mathcal{K})$, which is M. Let (t) denote $\Sigma_i\, b_i^2$ indeterminates. Then $\mathcal{K}(t) = \mathcal{K} \otimes_R R(t)$ gives a resolution of $M(t) = M \otimes_R R(t)$ over $R(t)$. We can make a change of basis in each $R(t)^{b_i}$ in which the change of basis matrix has as entries the appropriate b_i^2 indeterminates (so that the sets of entries of the n change of basis matrices are mutually disjoint). The matrix of each map is then multiplied on each side by a square matrix of indeterminates of appropriate size, and the two matrices of indeterminates used for a given map are disjoint (note that $R(t_{ij})$, $1 \leq i$, $j \leq n$, has an R-automorphism which takes the matrix (t_{ij}) to its inverse, so that the inverse of a matrix of "round brackets indeterminates" is just as "good" as the original matrix of indeterminates).

Generally speaking, if U is a p by q matrix over R, and (t) represents p^2+q^2 indeterminates, say (t_{ij}) and (t'_{ij}), then if S is faithfully flat over $R(t)$ we shall say $(t_{ij})U(t'_{ij})$ is in general position in S.

Lemma 2.6. _Let_ U* _be the result of putting an_ s _by_ r _matrix_ U _in general position,_ $s \geq r$, _and suppose that the ideal of_ r _by_ r _minors of_ U _has grade_ g . _Then the ideal of_ r _by_ r _minors of any_ r+g-1 _by_ r _submatrix of_ U* _has grade_ g . _(Rings are assumed Noetherian here.)_

Proof. The grade will be at most $(r+g-1)-r+1 = g$ (which is what it is for a matrix of indeterminates): the problem is to show that it is at least g . Faithfully flat extension does not change grade, and localization only increases grade. Multiplying on the right by the matrix of round brackets indeterminates does not change the ideal of minors of a given size. Hence, we might as well assume that we are only multiplying on the left by a matrix of square brackets indeterminates. The submatrix of U* then has the form TU , where T is an r+g-1 by r matrix of indeterminates, and if R was the original ring, we can work over the ring obtained by adjoining the entries t_{ij} of T to R . Let J be the ideal generated by the r by r minors of TU . If $G(J) < g$, then we can choose a prime Q of R[t] such that $D(R[t]_Q) < g$. If $Q \cap R = P$, we can replace R by R_P without altering the hypothesis $G(JR_P) < g$. Now, if the ideal J_0 of r by r minors of U is contained in P , then $J_0R[t] \subset Q$ and $G(J_0) = G(J_0R[t]) \leq G(Q) \leq D(R[t]_Q) < g$, a contradiction. Hence, we can assume that (R,P) is local and that some r by r minor of U has invertible determinant, i.e. is invertible as a matrix. Let V be the s by r matrix whose first r rows form an r by r identity matrix and whose last s-r rows are zero. Then U = AV where A is an s by s invertible matrix. Now, TA = T' is simply a new matrix of indeterminates over R , because A is invertible, and TU = T(AV) = T'V , which is an r+g-1 by r matrix of indeterminates (a submatrix of T'). It follows that the grade of the ideal of r by r minors of TU is g , as required.

We now continue the discussion of Conjecture (F) in dimension 2 and higher. Suppose that we are working in dimension n . Let M be a module of finite length and finite projective dimension (equivalently, let M be perfect, with pd M = n). Pass from R to R(t) for some large family of indeterminates (t) and assume that

$$0 \to R^{b_n} \to \dots \to R^{b_1} \to R^{b_0} \to 0$$

is a free minimal resolution of M with matrices in general position. Then $b_0, b_1, \dots, b_n$ are the Betti numbers of M and $\Sigma_i (-1)^i b_i = 0$, i.e. $b_0+b_2+\dots = b_1+b_3+\dots$. Let $\underline{i}(M) = (\Sigma_i b_i)/2 = b_0+b_2+\dots = b_1+b_3+\dots$. We call $\underline{i}(M)$ the <u>index</u> of M.

Let U be the b_1 by b_0 matrix of the first map, so that $M = \text{Coker } U$. Then $\text{Ann } M$ is power associated with the ideal of b_0 by b_0 minors of U, and the grade of this ideal must be n. Thus, $b_1 \geq b_0+n-1$. If equality holds, then M comes from a generic torsion module, and we are done. If not, $b_1 > b_0+n-1$. Let U_0 be the submatrix of U consisting of the first b_0+n-1 rows. The ideal of maximal minors of U_0 has grade n, by Lemma 2.6, and it follows that $M^* = \text{Coker } U_0$ is a perfect module of projective dimension n which comes from a generic torsion module. We get a short exact sequence $0 \to M_1 \to M^* \to M \to 0$ for a certain M_1. Then $\text{pd } M_1 \leq n$ while $\text{Ann } M_1 \supset \text{Ann } M^*$ and $G(\text{Ann } M_1) \geq n$. It follows that M_1 is perfect of projective dimension n as well. It is worth noting that from [3], we know the entire resolution of M^*.

Also note that M_1 is generated by the images of the last $b_1-(b_0+1)$ rows of U in the quotient of the row space of U by the submodule spanned by the first $b_1-(b_0+1)$ rows. These are minimal generators, since the rows of U are minimal generators for the row space. To give a relation on these generators is the same as to give a relation on all the rows of U and then project on the last $b_1-(b_0+1)$ coordinates. Hence, if V is the second matrix of the resolution of M, the rows of the submatrix of V consisting of the last $b_1-(b_0+1)$ columns generate (not necessarily minimally) the relations on the specified generators of M_1.

Now, let $r_i = \text{rank } (\text{Tor}_i(M^*,k) \to \text{Tor}_i(M,k))$, $0 \leq i \leq n$, where k is the residue class field of R.

Lemma 2.7. <u>The index</u> $i(M_1)$ <u>is</u> $\underline{i}(M) + (\underline{i}(M^*) - \Sigma_i r_i)$.

<u>Proof</u>. Let $T_i = \mathrm{Tor}_i(\ ,k)$ and consider the long exact sequence: $\to T_{i+1}(M) \to T_i(M_1) \to T_i(M^*) \to T_i(M) \to T_{i-1}(M_1) \to \ldots$. Let a_i, b_i, c_i be the ith Betti numbers of M_1, M, M^*, respectively. Then $a_i = \dim_k \mathrm{Tor}_i(M_1,k)$, and so forth. Since the rank of $T_i(M^*) \to T_i(M)$ is r_i , the rank of the preceding map is $c_i - r_i$, the rank of the following map is $b_i - r_i$, and it follows that $a_i = (b_{i+1} - r_{i+1}) + (c_i - r_i)$ for each i . Summing these equalities (and keeping in mind that $b_0 = r_0$) we get $2\underline{i}(M_1) = 2\underline{i}(M) + 2\underline{i}(M^*) - 2\Sigma_i r_i$, as required.

If it were true that $\Sigma_i r_i > \underline{i}(M^*)$ always, we would be able to show that the class of M in $F(R)$ is trivial by induction on $\underline{i}(M)$: the class of M^* is trivial, and $\underline{i}(M_1)$ would be less than $\underline{i}(M)$. The inequality $\Sigma_i r_i > \underline{i}(M^*)$ does hold in dimension 2 (we shall prove something quite a bit stronger), but in dimension 3 equality is possible.

In dimension 2, $b_1 = b_2 + b_0$, and the long exact sequence for Tor takes the form:

$$0 \to T_2(M_1) \to k \xrightarrow{f} k^{b_2} \to T_1(M_1) \xrightarrow{g} k^{b_0+1} \to$$

$$k^{b_1} \to k^{b_1-(b_0+1)} \to k^{b_0} \to k^{b_0} \to 0$$

Since $T_2(M_1) \neq 0$, $T_2(M_1) \cong k$, and f is 0 . Working backward from the right hand side, it is easy to see that g is 0 . Thus, M_1 has Betti numbers $b_1 - (b_0+1) = b_2 - 1$, b_2, and 1 , and comes from a generic torsion module. Moreover, its matrix is exactly the submatrix of V consisting of the last $b_2 - 1$ columns. We have shown:

Corollary 2.8. <u>Let</u> E <u>be a perfect module of projective dimension</u> 2 <u>over a local ring</u> R . <u>Then for some</u> (t) , $E(t)$ <u>is a quotient of a perfect module arising from a generic torsion module by a perfect submodule arising from a generic torsion module, both of projective dimension</u> 2.

In dimension 3, this technique seems to fail. However, one can still show that $\Sigma_i r_i \geq \underline{i}(M^*)$, so that passing from M to M_1 can never increase the index. For the long exact sequence takes the form:

$$0 \to k^{a_3} \to k^{b_0} \to k^{b_3} \to k^{a_2} \to k^{b_0+2} \to k^{b_2} \to k^{a_1}$$

$$\xrightarrow{f} k^{b_0+2} \to k^{b_1} \to k^{b_1-b_0-2} \to k^{b_0} \to k^{b_0} \to 0$$

and it is easy to see that f must be 0 . Now, $\underline{i}(M^*) = 2b_0+2$, and we can see that $r_0 = b_0$ and $r_1 = b_0+2$. Thus, $\underline{i}(M_1) < \underline{i}(M)$ unless $r_3 = r_2 = 0$, in which case we get equality. In the case $r_3 = r_2 = 0$, it is easy to see that $a_3 = b_0$, $a_2 = b_3+b_0+2$, and $a_1 = b_2$, in addition to $a_0 = b_1-b_0-2$, so that the new Betti numbers are completely determined in this case. This situation can occur, which indicates the need for more g-modules besides those coming from generic torsion modules in order to generate the stable Grothendieck group. This is not unexpected.

We next want to prove a result which yields Serre's conjecture in dimension ≤ 4 . However, we first want to note the following: let M and N be perfect modules over a C-M local ring (R,P) such that $\ell(M\otimes N) < \infty$. Let i be the biggest integer such that $\mathrm{Tor}_i(M,N) \neq 0$. Then $i = \dim R - \dim M - \dim N$. To see this, we first note that either from the main result of Buchsbaum and Eisenbud [2] or from [8], $i = \mathrm{pd}\, M - D(\mathrm{Ann}\, M, N)$.
But $\mathrm{pd}\, M - D(\mathrm{Ann}\, M, N) = D(R) - D(M) - D(\mathrm{Ann}\, M + \mathrm{Ann}\, N, N) =$
$D(R) - \dim M - D(P,N) = D(R) - \dim M - \dim N$, and $D(R) = \dim R$. Hence, if $\dim M + \dim N = \dim R$, there is only one non-vanishing Tor and $e(M,N) = \ell(M\otimes N)$, while if $\dim M + \dim N = \dim R - 1$ there are only two nonvanishing Tor's and $e(M,N) = \ell(M\otimes N) - \ell(\mathrm{Tor}_1(M,N))$.

Theorem 2.9. _Let_ (R,P) _be a regular local ring of dimension_ n . _Then the following two conditions are equivalent:_

1) _For any two finitely generated modules_ M, N _such that_ $\dim M + \dim N < n$ _and_ $\ell(M\otimes N) < \infty$, $e(M,N) = 0$.

2) <u>For any two perfect modules</u> M, N <u>such that</u> dim M + dim N = n-1 <u>and</u> $\ell(M \otimes N) < \infty$, e(M,N) = 0 , i.e. $\ell(M \otimes N) = \ell(\mathrm{Tor}_1(M,N))$.

<u>If Conjecture</u> (F) <u>holds for complete intersections in dimension</u> $\leq [(n+1)/2]$, <u>then</u> 1) <u>and</u> 2) <u>do hold for</u> R . <u>If</u> 1) <u>and</u> 2) <u>hold for</u> R <u>and</u> R <u>has abundant Cohen-Macaulay modules, then Conjecture</u> (C) <u>holds for</u> R .

<u>Proof</u>. We first prove the equivalence of 1) and 2) . Evidently, what we must show is that if 2) holds, then 1) holds. We proceed by induction on t = (pd M - G(Ann M)) + (pd N - G(Ann N)) + (dim R - dim M - dim N) . The case t = 1 is the hypothesis 2) . Now assume the result for smaller t , and assume that t > 1 . Let I, J be Ann M , Ann N , respectively. We first claim that there is a maximal R-sequence in I which has an initial segment that is a maximal N-sequence. To show this it is only necessary to show that dim N = (D(N)) = D(I,N) and that $G(I) \geq D(I,N)$, since we can then construct an appropriate sequence by avoiding the union of two sets of primes that must be avoided at each stage. But D(N) = D(P,N) = D(I+J,N) = D(I,N) , since JN = 0 , and $D(I,N) \leq \mathrm{ht}\ I = G(I)$, since R is regular and therefore C-M. Let G(I) = m and dim N = k , and let $r_1, \ldots, r_m$ be the required sequence. By construction, $N \otimes (R/(r_1, \ldots, r_k))$ has finite length, so that this certainly holds for $N \otimes (R/(r_1, \ldots, r_m))$. We can map a direct sum of copies of $M^* = R/(r_1, \ldots, r_m)$ onto M , since $(r_1, \ldots, r_m) \subset I$, and this leads to a short exact sequence: $0 \to M_1 \to M^{*s} \to M \to 0$. Now, M^{*s} is g-perfect, $\ell(M^{*s} \otimes N) < \infty$, and $\dim M^{*s} = \dim M$, since both are n - m . Therefore, $e(M^{*s},N) = 0$, and e(M,N) vanishes if and only if $e(M_1,N)$ vanishes. Note that M_1 is annihilated by $\mathrm{Ann}\ M^{*s} = (r_1, \ldots, r_m)$, so that $G(\mathrm{Ann}\ M) \geq m$, and if $\mathrm{pd}\ M \geq m+1$, then $\mathrm{pd}\ M_1 = \mathrm{pd}\ M - 1$, so that the integer t for the pair (M_1,N) is at least one less than what it was for (M,N) . This shows that we can assume that pd M = m = G(Ann M) , so that M is perfect, and by similar reasoning we can also assume that N is perfect. Hence, we must be in the case where t = dim R - dim M - dim N and t > 1 . Assume, say, that dim M < dim R . We must have that m > dim N , since dim N < dim R - dim M = m . Since $k \leq m-1$, we can let $M^* = R/(r_1, \ldots, r_{m-1})$, map M^{*s} onto M for a

suitable s , and we get a short exact sequence: $0 \to M_1 \to M^{*s} \to M \to 0$. Since $\text{pd}\, M = m$ and $\text{pd}\, M^{*s} = m-1$, $\text{pd}\, M_1 = m-1$. Moreover, $G(\text{Ann}\, M_1) \geq G(\text{Ann}\, M^{*s}) = m-1$, and it follows that M_1 is perfect of projective dimension $m-1$ and dimension $\dim M + 1$. From the fact that $k \leq m-1$ we still know that the tensor products involved have finite length, and $\dim M^{*s} + \dim N = \dim M + \dim N + 1 < \dim R$ because $t > 1$, so that from the fact that M^{*s} is g-perfect we can still conclude that $e(M^{*s}, N) = 0$ and that $e(M,N)$ vanishes if and only if $e(M_1,N)$ vanishes. Since the integer t for the pair (M_1,N) is precisely one less than what it was for the pair (M,N) , we are done.

We have demonstrated the equivalence of 1) and 2) . Now assume that (F) holds for complete intersections of dimension $\leq [(n+1)/2]$. We must show that 2) holds for R of dimension n . But since $\dim M + \dim N = n-1$, we know that for one of these two modules, say M , $\dim M \geq (n-1)/2$. By the same reasoning as before, there is an R-sequence in $J = \text{Ann}\, N$ which is also a maximal M-sequence; call this R-sequence $r_1, \ldots, r_d$, and let $S = R/(r_1, \ldots, r_d)$. Since $d \geq (n-1)/2$, $\dim S \leq (n+1)/2$, and conjecture (F) holds for S . But $e_R(M,N) = e_S(M \otimes_R S, N)$, and $M' = M \otimes_R S$ is a perfect S-module of finite length. We still have $\dim M' + \dim N = \dim S - 1$, since $\dim M' = \dim M - d$ and $\dim S = \dim R - d$. By Conjecture (F) we can choose (t) so large that $[M'(t)]$ lies in $g(S(t))$. But $e_{S(t)}(M'(t),N(t)) = e_S(M',N)$, and we know that $e_{S(t)}(M'(t),N(t))$ vanishes because $[M'(t)]$ lies in $g(S(t))$ and $e_{S(t)}(\ ,N(t))$ vanishes on the generators of $g(S(t))$.

It remains to prove that if 1) holds and R has abundant C-M modules, then (C) holds. We must show that if $\dim M + \dim N = \dim R$, then $e(M,N) > 0$. This reduces to the case where M, N are prime cyclic, as usual, say $M = R/Q$, $N = R/Q'$. Let T, T' be Cohen-Macaulay modules over R having annihilators Q, Q' respectively. In a prime filtration for T or T' , Q or Q' will occur a positive number of times, say b or b' , and the other primes will be of strictly higher height, and their factors will contribute 0 to the multiplicity. It follows that $e(T,T') = bb'e(R/Q,R/Q')$, and it suffices to show that $e(T,T') > 0$. But since T and T' are perfect and

$\dim T + \dim T' = \dim R$, there is only one non-vanishing Tor , and $e(T,T') = \ell(T \otimes T') > 0$, as required.

Corollary 2.10. If Conjecture (E) holds (in dimension n) and Conjecture (F) holds for complete intersections (in dimension $\leq [(n+1)/2]$) , then Conjecture (C) holds (in dimension n) .

Proof. Conjecture (C) can be reduced to the complete case. But then R has abundant C-M modules, by Conjecture (E), and the result is immediate from Theorem 2.9.

Corollary 2.11. Conjecture (C) holds for regular local rings of dimension ≤ 4.

Proof. We can reduce to the complete case. Since we know (F) in dimension ≤ 2 , and since $2 \leq [(4+1)/2]$, it suffices to show that complete regular local rings in dimension ≤ 4 have abundant C-M modules. If the prime Q has height ≤ 1 , R/Q itself is C-M . If the prime Q has coheight ≤ 1 , this is true again. The only remaining case is where $\dim R = 4$ and $\operatorname{ht} Q = 2$. But Conjecture (E) is known for complete local domains of dimension 2 (see the next section).

We conclude this section by noting one further consequence of (E) which is, in fact, proved to be equivalent to (A') in Peskine and Szpiro [14], and is known precisely when (A') is known: the zerodivisor conjecture.

Conjecture (D). If M is a finitely generated module of finite projective dimension over a local ring R and $r_1, \ldots, r_k$ is an M-sequence , then $r_1, \ldots, r_k$ is an R-sequence.

3. THE TWO-DIMENSIONAL CASE

We want to give several proofs of (E) when the dimension of the ring is 2. We can assume that we are dealing with a complete local domain.

First proof. Let $M = R'$ be the integral closure of R . Since R' is an integrally closed local domain of dimension 2 , $D_{R'}(R') = 2$. But then

$D_R(R') = D_{R'}(R') = 2$. This is essentially the technique of Levin and Vasconcelos [12] for proving Bass' conjecture in the $D(R) = 1$ case. Note also that Bass' conjecture is the same as the conjecture "IV implies III" which is raised as a question (not conjectured) on p. 162 of [10].

<u>Second proof</u>. Since the ring S is a complete local domain, S is a module-finite extension of a complete regular local ring R . Then $\mathrm{Hom}_R(S,R)$ has an S-module structure and is free as an R-module. For $D_R(S) \geq 1$ implies that $\mathrm{pd}_R S \leq 1$, and if $0 \to R^n \to R^m \to S \to 0$ is exact, then we get:

$$(*)\ 0 \to \mathrm{Hom}_R(S,R) \to R^n \to R^m \to \mathrm{Ext}^1_R(S,R) \to 0$$

is exact, and since $\mathrm{Hom}_R(S,R)$ is a second module of syzygies of $\mathrm{Ext}^1_R(S,R)$ over a 2-dimensional regular local ring it must be free.

One can also see directly that $D(\mathrm{Hom}_R(S,R)) \geq 2$, for if r_1, r_2 is an R-sequence and $r_2 g = r_1 f$, $f, g \in \mathrm{Hom}(S,R)$, then for each s in S , $r_2 g(s) = r_1 f(s)$, and so r_2 divides $f(s)$ for every s ; but then r_2 divides f .

<u>Third proof</u>. This proof is a variation on the same basic idea as the second proof: applying a functor to an S-module. The functor used here is similar to the one used above. Let E be a finitely generated module over S , where (S,Q) is local and complete, and let $T^i = D(H^i_Q)$, where H^i_Q is local cohomology and $D = \mathrm{Hom}_S(\ , E(S/Q))$ ($E(S/Q)$ is the injective hull over S of S/Q). See Grothendieck-Hartshorne [7], pp. 87 and 88. (Note: if we write $S = R/I$, where R is regular of dimension m , then $T^i(E) \cong \mathrm{Ext}^{m-i}_R(E,R)$; one does not want this as a definition, however, because it seems to depend on the choice of R .) Let $n = \dim E$, and suppose s (respectively, s_1, s_2) is part of a system of parameters for E . Then s (respectively, s_1, s_2) is a $T^n(E)$-sequence. Thus, if $\dim S = 2$ and E is any 2-dimensional finitely generated S-module, $T^2(E)$ is a maximal C-M module over S .

To see that the statements above hold, note that since $\dim E/sE = n-1$,

$T^n(E/sE) = 0$. Consider the exact sequence $0 \to E'' \to E \to E' \to 0$, where $E'' = \cup_n \mathrm{Ann}_E s^n$. Since $\dim E'' < \dim E$, $T^n(E'') = 0$, and it follows from the long exact sequence for T that $T^n(E') \cong T^n(E)$. But s is not a zerodivisor on E' , and the exact sequence $0 \to E' \xrightarrow{s} E'/sE' \to 0$ yields $0 \to T^n(E'/sE') \to T^n(E') \xrightarrow{s} T^n(E') \to \ldots$. Since $\dim E'/sE' < n$, $T^n(E'/sE') = 0$, and s is a nonzerodivisor on $T^n(E') \cong T^n(E)$, as required. Now suppose $s = s_1, s_2$ is part of a system of parameters for E . Define E' as above. We must show that s_2 is not a zerodivisor on $T^n(E)/sT^n(E) \cong T^n(E')/sT^n(E')$. But we have an exact sequence $0 \to T^n(E') \xrightarrow{s} T^n(E') \to T^{n-1}(E'/sE') \to \ldots$, and so we have an injection $0 \to T^n(E)/sT^n(E) \to T^{n-1}(E'/sE')$. Since $\dim E'/sE' = n-1$ and s_2 is part of a system of parameters for E'/sE' , s_2 is not a zerodivisor on $T^{n-1}(E'/sE')$, and hence not a zerodivisor on $T^n(E)/sT^n(E)$. (This result does not extend to systems of parameters of length three: let S be a regular local ring, and E a torsion-free reflexive module which is not free. Then $T^3(E) = \mathrm{Hom}_S(E,S)$ is not free and has depth < 3 .)

Fourth proof. Let Q be a height one prime of a complete local domain S of dimension ≥ 2 . Let $Q^{(n)} = Q^n S_Q \cap S$, as usual, the nth symbolic power. Then for all sufficiently large n , $D(Q^{(n)})$ (not $G(Q^{(n)})$!!) is at least 2. We shall prove this in §5 . It is the circle of ideas involved in this proof which we feel is most likely to generalize to higher dimensions.

4. APPROACHES TO HIGHER DIMENSION

In this section we consider several techniques which might be used to prove Conjecture (E) in higher dimension. One technique is reserved for §5 , however.

Let R be a complete regular local ring. Consider the smallest class $\mathcal{C}$ of module-finite extension domains of R such that 1) R is in $\mathcal{C}$, 2) if S is in $\mathcal{C}$ and $R \subset S_0 \subset S$, then S_0 is in $\mathcal{C}$, and 3) if S is in $\mathcal{C}$ and T is an S-free module finite extension domain of S , then T is in $\mathcal{C}$.

We shall call the R-algebras in $\mathcal{C}$ the accessible R-algebras. It is quite trivial that the class $\mathcal{G}$ of module-finite extension domains of R which possess maximal C-M modules satisfies 1), 2), and 3). In case 2), if M is a maximal C-M module over S, it is also a maximal C-M module over S_0. In case 3), if M is a maximal C-M module over S, then $M \otimes_S T$ is a maximal C-M module over T. Thus, accessible R-algebras have maximal C-M modules.

In dimensions 1 and 2 every module-finite extension domain of R is accessible: if $R \subset S$, the integral closure S' of R is R-free and hence in $\mathcal{G}$; but then its subalgebra S is in $\mathcal{G}$. This gives essentially the same proof as the first proof of (E) in dimension 2 of the preceding section.

Thus, it is natural to ask in higher dimensions:

Question 4.1. Is every module-finite extension domain of a complete regular local ring accessible?

An affirmative answer would prove (E). Note that the accessible R-algebras are all R-split (i.e. will contain R as a direct summand as R-modules), since the class of R-split extensions satisfies the conditions 2), 3). Thus, we cannot have an affirmative answer to Question 4.1 unless every module-finite extension domain of R is R-split when R is a complete regular local ring. This is true: see [9].

Note, however, that there are 3-dimensional complete local rings which have no module-finite C-M extension algebra. In fact, let R be an integrally closed algebro-geometric local domain over a field of characteristic 0, and suppose $\dim R = 3$ and R is not C-M. We give such an example in §5. Then the completion S of R is an integrally closed 3-dimensional complete local ring which is not C-M. Suppose T were a module-finite C-M extension algebra of S. Because S contains the rationals we can retract T onto S by means of a modified trace map (cf. [9]). Since S is a direct summand of T as S-modules, we have $D_T(T) = D_S(T) \le D_S(S) < 3$.

A different approach to the higher dimensional case involves taking the point of view of Conjecture (E'''). Let S be a module-finite extension of R, let

$s_1, \ldots, s_m$ be a set of generators for S as an R-module (for convenience, assume $s_1 = 1$), and let $(r_{i1}, \ldots, r_{im})$, $1 \leq i \leq t$, generate the module of relations on $s_1, \ldots, s_m$ over R. For each i, j we can write $s_i s_j = \Sigma_{k=1}^{m} r_{ijk} s_k$, $1 \leq i,j \leq m$.

In order to map S homomorphically into $\mathfrak{M}_n(R)$ we must find values in R for mn^2 indeterminates x_{ijk} such that if X_k denotes the n by n matrix (x_{ijk}), $1 \leq k \leq m$, then the values for the indeterminates satisfy the matrix equations:

$$X_1 = \text{the } n \text{ by } n \text{ identity matrix.}$$

$$(\xi_n) \qquad \Sigma_j r_{ij} X_j = 0, \quad 1 \leq i \leq t .$$

$$X_i X_j = \Sigma_{k=1}^{m} r_{ijk} X_k, \quad 1 \leq i,j \leq m .$$

We want to know whether, for some n, the equations (ξ_n) have a solution in R.

Let $r_1, \ldots, r_d$ be a regular system of parameters for R. Suppose that for fixed n and for every integer q, we can solve (ξ_n) modulo $I_q = (r_1^q, \ldots, r_d^q)$. In certain good cases (e.g. if R has an uncountable algebraically closed residue class field) it is possible to show that a system of equations has a solution in R if and only if it has a solution modulo every power of the maximal ideal or, equivalently, modulo every I_q. But to solve (ξ_n) modulo I_q is the same as to find an (S/I_qS)-module which is free of rank n as an (R/I_q)-module. Hence, if R is a complete regular local ring with an uncountable algebraically closed residue class field and S is a module-finite extension domain of R, then S possesses a maximal C-M module if and only if for some integer n and every integer q, S/I_qS possesses a finitely generated module which is free of rank n as an (R/I_q)-module.

Since R/I_q is a 0-dimensional Gorenstein local ring, i.e. a self-injective local ring, it might be tempting to conjecture that if R is a 0-dimensional Gorenstein local ring and S is a module-finite local extension, then some S-module is R-free, but this is false. E.g. let $S = k[x] = k[X]/(X^3)$, where k is a field and X is an indeterminate, and let $R = k[x^2]$. It is quite easy to see that no S-module is R-free.

Quite generally, if $(R,P) \subset (S,Q)$ are 0-dimensional local rings with the same socle $yR = yS$ (which implies that $R/P \cong S/Q$) , and $R \neq S$, then no nonzero S-module is R-free. For if E is an S-module which is R-free, $E \neq 0$, and $e_1, \ldots, e_t$ are minimal generators for E as an R-module, then they are minimal generators for E as an S-module: in fact, $QE = PE$. For if u is in $QE - PE$, then $yu = 0$, and if $u = \Sigma_i r_i e_i$ and not all the r_i are in P , then $yu = \Sigma_i (yr_i)e_i = 0$ gives an R-relation on the e_i . Now, since $ye_1 \neq 0$, $\mathrm{Ann}_S e_1 = 0$, and $S \cong Se_1$ is a submodule of E and therefore a direct summand as S-modules (since S is S-injective). But then $PS = QS = Q$, since $PE = QE$. We also know that $R/P = S/Q$, and it follows that $R = S$.

This argument yields the following:

Proposition 4.2. Let (R,P) be a complete regular local ring with R/P algebraically closed and let S be a module-finite domain extension, $S \neq R$. Suppose that S possesses a maximal C-M module. Then for each irreducible P-primary ideal I of R , there is an ideal J of S properly larger than IS such that J lies over I .

Proof. S/IS is a local extension of R/I and R/I is Gorenstein. Hence, S/IS cannot have the same socle as R/I , and there is an ideal J/IS of S/IS disjoint from the socle of R/I . J is the required ideal.

Thus, if Conjecture (E) is true, the conclusion of (4.2) holds without the extra assumption on S . It would be interesting to know whether this is really true.

Another idea which might work in the higher dimensional case is to show that if S is a module-finite extension domain of a complete regular local ring R , then for some functor F from finitely generated R-modules to themselves and for some finitely generated S-module E , F(E) is R-free and $\neq 0$. For F(E) is still an S-module: in fact, it is a $\mathrm{Hom}_R(E,E)$-module. This general idea is utilized in the second proof of (E) in dimension 2 in the preceding section, and a variation (in which the functor is defined only on a subclass of finitely generated R-modules, namely, S-modules) is used in the third proof. We note, for example, that in dimension 3 it would be sufficient to find a torsion-free S-module E of homological dimension 1 over R such that $D_R(\mathrm{Ext}^1_R(E,R)) > 0$. For then, by the same exact sequence (*) given in the second proof of (E), we can conclude that $\mathrm{Hom}_R(E,R)$ is R-free.

We devote the next section to our remaining suggested approach to the higher-dimensional case. We conclude this section with the remark that, for certain purposes, adjoining round brackets indeterminates is harmless in looking for maximal C-M modules over a complete local domain.

In other words, if (S,P) is a complete local domain and (t) denotes several indeterminates, then if S(t) possesses a maximal C-M module, so does S. To see this, represent S as a module-finite domain extension of a complete local ring R. Then S(t) is module-finite over R(t), and we have a homomorphism $f: S(t) \to \mathcal{M}_n(R(t))$ for some n. We can choose a polynomial h in R[t] with some coefficient outside the maximal ideal of R such that $f(S) \subset \mathcal{M}_n(R[t, 1/h])$. We can choose an R-free module-finite extension domain (R',P') of R such that h does not vanish identically on R'/P', and then we can choose values for the variables t in R' such that h specializes to a unit of R'. This gives a homomorphism $R[t,1/h] \to R'$ which induces a homomorphism $S \to \mathcal{M}_n(R')$. But since R' is R-free we can view $R' \subset \mathcal{M}_m(R)$ for some m, and then we have a homomorphism $S \to \mathcal{M}_{nm}(R)$.

5. DEFICIENCIES AND SYMBOLIC POWERS

Throughout this section, (R,P) denotes a local ring and $p_1, \ldots, p_w$ denote generators for P. If $M \neq 0$ is a finitely generated R-module, we want to associate certain integers with M which we call the <u>deficiencies</u> of M (the definition depends on the choice of $p_1, \ldots, p_w$ but we shall ignore this in our notation). The 0th <u>deficiency</u> $\mathrm{def}^R_0 M = \mathrm{def}_0 M$ we define to be the smallest integer d such that $L(M) \cap P^d M = 0$, where $L(M) = H^0_P(M) = \{m \in M: P^t m = 0$ for some integer $t\} = \cup_t \operatorname{Ann}_M P^t$. To define the kth deficiency let t_{ij} be kw indeterminates, let $s_i = \Sigma_{j=1}^{w} t_{ij}p_j$, $1 \leq i \leq k$, and let $\operatorname{def}_k M = \operatorname{def}_0^{R(t)} M(t)/(s_1, \ldots, s_k)M(t)$.

It is trivial to see that the first nonvanishing deficiency of a nonzero module M occurs when $k = D(M)$, and we regard it as a measure of how far away M is from having depth one more than it actually does.

The significance of this notion for us here is a consequence of the following:

Theorem 5.1. Let M be a finitely generated module over (R,P) with $D(M) \geq k$, and let N be a nonzero submodule of M such that $D(M/N) \geq k$. Suppose that $N \subset P^{\operatorname{def}_k M} M$. Then $D(N) \geq k+1$.

Proof. The question of whether $D(N) \geq k+1$ is unaffected by tensoring with $R(t)$. Thus, we may adjoin kw indeterminates (t) as in the definition of kth deficiency, and we obtain a sequence $s_1, \ldots, s_k$ in "general position". It follows that $s_1, \ldots, s_k$ is a regular sequence on both $M(t)$ and $(M/N)(t) \cong M(t)/N(t)$, and hence on $N(t)$ as well, so that $D(N(t)) \geq k$. Let $'$ denote $\otimes_{R(t)} R(t)/(s_1, \ldots, s_k)$. Then it remains to show that $P(t)$ is not an associated prime of $N(t)'$. Since $0 \to N(t)' \to M(t)' \to (M(t)/N(t))' \to 0$ is exact, we have that $N(t) \cap (s_1, \ldots, s_k)M(t) = (s_1, \ldots, s_k)N(t)$. Suppose we have an element u of $N(t)$ such that $P(t)u \subseteq (s_1, \ldots, s_k)N(t)$. We must show that $u \in (s_1, \ldots, s_k)N(t) = N(t) \cap (s_1, \ldots, s_k)M(t)$, so it suffices to show that $u \in (s_1, \ldots, s_k)M(t)$. Working modulo $(s_1, \ldots, s_k)M(t)$ we see that $u' \in L(M(t)')$, since $P(t)u \subset (s_1, \ldots, s_k)N(t) \subset (s_1, \ldots, s_k)M(t)$, and $u' \in P(t)^{\operatorname{def}_k M} M(t)'$, since $u \in N \subset P^{\operatorname{def}_k M} M$. Hence,

$$u \in L(M(t)') \cap P(t)^{\operatorname{def}_0 M(t)'} M(t)' = 0 ,$$

i.e. $u \in (s_1, \ldots, s_k)M(t)$, as required.

We can now complete the argument in the fourth proof of (E) in dimension 2 given in §3 .

Corollary 5.2. *Let* (R,P) *be a complete local domain and* Q *any nonzero prime of coheight at least one. Then for all sufficiently large* m , $D(Q^{(m)}) \geq 2$.

Proof. Let $d = \text{def}_1 R$. Clearly, D(R) and $D(R/Q^{(m)})$ are at least one for all m , since any element of P - Q is a nonzerodivisor on $R/Q^{(m)}$. Hence, if m is so large that $Q^{(m)} \subset P^d$, we are done. But since R is a domain the intersection of the symbolic powers of Q is 0 , and since R is complete it follows that $Q^{(m)}$ is contained in a given power P^t of P for all sufficiently large m .

In fact, suppose that whenever M is an R-module and Q a prime of R we define $Q^{(m)}[M]$ to be the kernel of the natural map $M \to (M/Q^m M) \otimes R_Q$, so that $Q^{(m)}[M] = \{u \in M:$ for some a in R-P , $au \in P^m M\}$. Then we have:

Lemma 5.3. *Let* (R,P) *be a complete local domain and let* $Q \subset Q_1$, $Q \neq Q_1$ *be primes of* R . *Let* M *be a torsion-free* R-*module and let* t *be a given integer. Suppose either that:* 1) $Q_1 = P$, *or*

2) R_{Q_1} *is integrally closed, or*

3) R_{Q_1} *is analytically irreducible.*

Then for every sufficiently large integer m , $Q^{(m)}[M] \subset Q_1^{(t)}[M]$.

Proof. 1) implies 3) obviously, and 2) implies 3) by Corollary (37.8) of Nagata [13] , since R is module-finite over a complete regular local ring. By localizing, we can assume instead that (R,P) is analytically irreducible and $Q_1 = P$. Since M is torsion-free over R we can embed M in a free module $F = R^k$. By the Artin-Rees lemma we can choose an integer t* such that $P^t M \subset P^{t*}F$. Then it suffices to show that $Q^{(m)} \subset P^{t*}$ for large m , since $Q^{(m)}[M] \subset Q^{(m)}[F] \subset P^{t*}F$. Let S be the completion of R . To show that $Q^{(m)} \subset P^{t*}$ for large m , we only need to show that $\bigcap_m Q^{(m)}S = 0$. Since S is a domain, if Q* is any prime of S lying over Q , we have $\bigcap_m Q^{*(m)} = 0$, while $Q^{(m)} \subset Q^{*(m)}$, and we are done.

At this point we want to propose that a systematic study of deficiencies

$(\operatorname{def}_i Q^{(m)}[M])$ and *residual* deficiencies $(\operatorname{def}_i M/Q^{(m)}[M])$ of symbolic powers of primes on modules be made, because understanding the behavior of these functions as $m \to \infty$ could well lead to a proof of Conjecture (E).

The general idea is to look for filtrations $0 = M_0 \subset M_1 \subset \ldots \subset M_n = M$, where the inclusions are proper, and M is a module of dimension n over a complete local domain (R,P) such that the filtration satisfies the conditions:

1) $D(M/M_i) \geq 1$, $0 \leq i \leq n-1$ (so that $D(M_{i+1}/M_i) \geq 1$ for each i as well), and

2) $M_{i+1}/M_i \subset P^{\operatorname{def}_{n-i-1}(M_{i+2}/M_i)} M_{i+1}/M_i$, $0 \leq i \leq n-2$.

If one has such a filtration one can show by induction on $n-i$ that $D(M_{i+1}/M_i) \geq n-i$, $0 \leq i \leq n-1$, so that, in particular, $D(M_1) \geq n$.

We shall now restrict attention to dimension 3 for a while to see what might be enough. Let (R,P) be a complete local domain of dimension 3 and let M be a torsion free R-module of finite type. If for each i we can find $N_i \subset P^i M$ such that $D(M/N_i) \geq 2$, then eventually the N_i have depth 3, by Theorem 5.1. In fact, by a first application of Theorem 5.1, the facts that $D(M/N_i) \geq 1$ and $N_i \subset P^{\operatorname{def}_1 M} M$ imply that $D(N_i) \geq 2$, and then since $D(M/N_i)$ and $D(N_i)$ are both ≥ 2, we have that $D(M) \geq 2$. We can then apply Theorem 5.1 again: $D(M) \geq 2$, $D(M/N_i) \geq 2$, and $i \geq \operatorname{def}_2 M$ taken together imply that $D(N_i) \geq 3$.

However, if we are taking the sequence N_i to be a subsequence of $Q^{(m)}[M]$, where Q is a height one prime of R, we do not need to know quite as much: instead of requiring $D(M/N_i) \geq 2$, so that $\operatorname{def}_1(M/N_i) = 0$, we only need to require that $\operatorname{def}_1(M/N_i)$ be bounded as $i \to \infty$. However, we shall need to assume that R_Q is a discrete valuation ring. This is true for all but finitely many height one primes in a complete local domain since the singular locus is closed.

Proposition 5.4. *Let* (R,P) *be a complete local domain of dimension* 3. *Let* Q *be a height one prime of* R *such that* R_Q *is a discrete valuation ring. Let* M *be a finitely generated torsion-free module over* R. *Suppose that* $\operatorname{def}_1(M/Q^{(i)}[M])$ *is bounded as* $i \to \infty$. *Then for all sufficiently large* i,

$D(Q^{(i)}[M]) = 3$.

<u>Proof</u>. Since R is complete and the singular locus is closed we can choose a height two prime Q_1 of R such that $Q \subset Q_1$ and R_{Q_1} is regular. Let k bound $\mathrm{def}_1(M/Q^{(i)}[M])$ and choose $j' \geq \max\{k, \mathrm{def}_1 M\}$. Choose j (by Lemma 5.3) so large that $M_1 = Q_1^{(j)}[M] \subset P^{j'}M$. Let $d = \mathrm{def}_2 M_1$. Choose i so large that $Q^{(i)}[M] \subset P^d M_1$ (which is possible from Lemma 5.3 and the Artin-Rees lemma). Since $\mathrm{def}_1(M/Q^{(i)}[M]) \leq k$, and $M_1/Q^{(i)}[M] \subset P^k(M/Q^{(i)}[M])$, it follows that $D(M_1/Q^{(i)}[M]) \geq 2$. Since $Q^{(i)}[M] \subset P^d M_1$, we have that $D(Q^{(i)}[M]) \geq 3$, as required. (We have found a good filtration.)

Remark 5.5. It is possible to bound deficiencies in a more "functorial" way. Let M be a finitely generated R-module, (R,P) local. Let $T = R + Pz + P^2z^2 + \ldots = R[Pz] \subset R[z]$, where z is an indeterminate over R . Then there is a functor F from R-modules to graded T-modules: $F(M) = M + PMz + P^2Mz^2 + \ldots \subset M[z] = M \otimes_R R[z]$, i.e. $F(M) = \mathrm{Im}(M \otimes_R T \to M \otimes_R R[z])$. Then

$$\mathrm{def}_0(M) \leq \ell(H^0_{PT}(F(M)) ,$$

since $H^0_{PT}(F(M)) = \Sigma_i\ (H^0_P(M) \cap P^iM)z^i$ and $H^0_P(M) \cap P^iM$ makes a nonzero contribution to the described length for each i , $0 \leq i \leq \mathrm{def}_0 M$. It follows at once that

$$\mathrm{def}_i(M) \leq \ell(H^0_{PT(t)}(F(M(t) \otimes_{R(t)} (R(t)/(s_1, \ldots, s_i)))) .$$

We next want to make estimates of the behavior of $\mathrm{def}_1(M/Q^{(i)}[M])$ under good conditions.

Proposition 5.6. <u>Let</u> (R,P) <u>be a complete local domain of dimension</u> 3, <u>let</u> M <u>be a finitely generated torsion-free</u> R-<u>module, and let</u> Q <u>be a height one prime of</u> R <u>such that</u> R_Q <u>is discrete valuation ring and such that</u> $M/Q^{(1)}[M]$ <u>has depth</u> 2. <u>Then there is a constant</u> c <u>such that</u> $\mathrm{def}_1(M/Q^{(i)}[M]) \leq ci$ <u>for</u>

<u>all</u> i .

<u>Proof</u>. Choose $x \in Q$ such that $R_q = xR_Q$ and choose y in $P - Q$ such that $yQ^{(1)}[M] = xM$. Let s be a general linear combination of generators of P in $R(t)$, as in the definition of def_1. Then y, s is a system of parameters for $R(t)/QR(t)$ and $(M/Q^{(1)}[M]) \otimes R(t) = M(t)/Q(t)^{(1)}[M(t)]$.

We shall begin by proving by induction on i that
$w \in H_i = H^0_{P(t)}(M(t)/(Q(t)^{(i)}[M(t)]+sM(t)))$ if and only if
$y^{i-1}w \in E_i = Q(t)^{(i)}[M(t)]+sM(t)$. On the one hand, if $y^{i-1}w$ is in E_i then $(y^{i-1}, Q^{(i)}, s)w \subset E_i$, and $\text{Rad}\,(y^{i-1}, Q^{(i)}, s) \supset \text{Rad}\,(Q, y, s) = P(t)$.

On the other hand, assume $w \in H_i$. If $i = 1$, we have that $y^j w + sv \in E_1$. But since $M(t)/Q(t)^{(1)}[M(t)]$ is C-M of depth 2 over $R(t)/Q(t)$ and since y, s is a system of parameters for $R(t)/Q(t)$, from the fact that $y^j w + sv \in Q(t)^{(1)}[M(t)]$ we can deduce that $w = y^{1-1}w$ is in $Q(t)^{(1)}[M(t)]+sM(t)$.

Now suppose $i > 1$, and $y^j w + sv \in Q(t)^{(i)}[M(t)]$. Since $Q(t)^{(i)}[M(t)] \subset Q(t)^{(1)}[M(t)]$, it follows that w is in $Q(t)^{(1)}[M(t)]+sM(t)$. Suppose $w = w' + sv'$, where w' is in $Q(t)^{(1)}[M(t)]$ and v' is in $M(t)$. We then obtain $y^j w' + sv'' \in Q(t)^{(i)}[M(t)]$, where v'' then must be in $Q(t)^{(1)}[M(t)]$ as well. Then $y^j(yw') + s(yv'')$ is in $yQ(t)^{(i)}[M(t)]$ and $yw' = xw^*$, $yv'' = xv^*$. It follows that $y(y^j w' + sv'') = x(y^j w^* + sv^*)$ and so $y^j w^* + sv^*$ is in $Q(t)^{(i-1)}[M(t)]$. By the induction hypothesis, we then have that $y^{i-2}w^*$ is in E_{i-1}. Then $y^{i-1}w = y^{i-1}(w' + sv') =$
$y^{i-1}w' + s(y^{i-1}v') = y^{i-2}xw^* + s(y^{i-1}v') \in xE_{i-1} + sM(t) \subset E_i$, as required.

Now choose d such that $P(t)^d \subset (Q(t), s, y)$. Let $c = 2d$. We need only show that $H_i \cap P(t)^{ci}M(t) \subset E_i$. Suppose that w is in $H_i \cap P(t)^{2di}M(t)$. Then w is in $(Q, s, y)^{2i}M(t) \subset Q^iM(t) + sM(t) + y^iM(t)$. Since $Q^iM(t) + sM(t) = E_i \subset H_i$, $w = w' + h$ where $w' \in y^iM(t)$ and $h \in E_i \subset H_i$, so that $w' \in H_i \cap y^iM(t)$, say $w' = y^i w''$. Choose j such that $y^j w' \in E_i$. Then $y^{j+i}w'' \in E_i$, so that $y^{i-1}w'' \in E_i$. But then $w' = y^i w'' \in E_i$, so that $w = w' + h \in E_i + E_i = E_i$, as required.

This result falls short of what we would like to have: some additional condition is needed to guarantee boundedness of the residual deficiency. But the conditions of the hypothesis are very easy to satisfy, so that it might well be possible to meet extra conditions.

The following lemma shows how to construct situations which satisfy the hypothesis of Proposition 5.6. D. Eisenbud suggested trying this type of construction to me.

Lemma 5.7. *Let (R,P) be a local domain, and let Q be a height one prime such that R_Q is a discrete valuation ring. Let E be a finitely generated torsion-free module over R/Q. Then there is a finitely generated torsion-free R-module M such that $M/Q^{(1)}[M] \cong E$.*

Proof. The localization of E at the prime 0 of R/Q is a finite-dimensional vector space over the field of fractions K of R/Q. We can embed E in a free (R/Q)-module $F = (R/Q)^t$ in such a way that the inclusion $0 \to E \to F$ becomes an isomorphism when we localize at the prime 0 of R/Q. We can therefore represent E as $\operatorname{Im} f$, where $f\colon (R/Q)^p \to (R/Q)^t = F$ in such a way that $(R/Q)^p$ maps onto the image of E in F. We can lift f to a map g of free R-modules: $g\colon R^p \to R^t$. Let $M = g(R^p)$. $M \subset R^t$ is torsion-free. Since the diagram

$$\begin{array}{ccccc} R^p & \xrightarrow{\text{onto}} & M & \hookrightarrow & R^t \\ {\scriptstyle\text{onto}}\ \big\downarrow u & & \big\downarrow \phi & & v \big\downarrow\ {\scriptstyle\text{onto}} \\ (R/Q)^p & \xrightarrow{\text{onto}} & E & \hookrightarrow & (R/Q)^t \end{array}$$

commutes $E = f(u(R^p)) = v(g(R^p)) = v(M)$, and there is an induced map ϕ of M onto E. It remains to see that the kernel of this map is precisely $Q^{(1)}[M]$, and for this purpose it suffices to compute the kernel after localizing at Q. We may therefore assume that (R,Q) is a discrete valuation ring and that $R/Q = K$, so that the inclusion of E in $(R/Q)^t$ is an isomorphism. Then M is torsion-free and consequently free, and we are trying to show that the kernel of ϕ is QM. It is clear that QM is contained in the kernel, and we know that the map of M/QM to E induced by ϕ is surjective. It suffices to show that this map is an isomorphism, or, equivalently, that $\dim_K(M/QM) = \dim_K E = t$. Since ϕ is surjective, $\dim_K(M/QM) \geq t$, and since $M \subset R^t$, $\dim_K(M/QM) = \text{rank } M \leq t$, and we are done.

Now, given a complete local domain (R,P) of dimension 3 we can choose a height one prime Q such that R_Q is a discrete valuation ring. We can construct a maximal C-M module E over R/Q, and we can then construct a torsion-free module over R such that $M/Q^{(1)}[M] \cong E$. This verifies the statement that the hypothesis of Proposition 5.6 is easily satisfied.

To see that some extra condition really is needed to guarantee that $\text{def}_1(M/Q^{(i)}[M])$ is bounded, not merely $O(i)$, we consider the following:

Example 5.8. Let $R = k[[x,y,z]]$, where k is a field. Let M be the ideal $(x,yz) \subset R$, and let $Q = zR$. Then $M/Q^{(1)}[M] \cong x(R/zR) \cong R/Q$ has depth 2, but $Q^{(i)}[M] = (xz^i, yz^i) \cong (x,y)$ has depth 2 for all i, which shows that $\text{def}_1(M/Q^{(i)}[M])$ is not bounded.

Despite this, there are interesting cases where the idea of Proposition 5.4 succeeds. We conclude this paper with one class of such examples.

Example 5.9. Let k be an algebraically closed field of characteristic 0. Let f be an irreducible homogeneous polynomial in $k[X_1, X_2, X_3]$ such that X_1, X_2, f is a homogeneous system of parameters but $X_3^2 \notin (X_1, X_2, f)$, and such that the partial derivatives of f with respect to X_1, X_2, X_3 vanish simultaneously only at $(0,0,0)$. (One example is $f = X_1^3 + X_2^3 + X_3^3$.) These conditions guarantee that $R_1 = k[X_1, X_2, X_3]/(f) = k[x_1, x_2, x_3]$ is integrally closed but not proper in the sense of Chow [4], p. 817, and it follows from the Theorem on p. 818 of [4] that the Segre product of R_1 with the polynomial ring $R_2 = k[y_1, y_2]$ is not Cohen-Macaulay. This Segre product $R = k[x_i y_j\colon\ i = 1, 2, 3 \text{ and } j = 1, 2] \subset R_1[y_1, y_2]$ is then an integrally closed ring of dimension 3 which is not Cohen-Macaulay. We shall construct a depth 3 module using the ideas of this section. (We could localize at the homogeneous maximal ideal and complete without essentially changing the situation, but it will be simpler to continue in the graded case.)

Let $Q = y_1R_1[y_1,y_2] \cap R$. Then $R/Q \subset R_1[y_2]$ is a domain; in fact, $R/Q \cong k[x_1y_2,x_2y_2,x_3y_2] \cong R_1$ is two-dimensional, and Q is a height one prime of R . We shall show that $Q^{(i)}$ is a C-M module for large i . In fact, we shall show that $D(R/Q^{(i)}) = 2$ for all i , which certainly suffices. To see this, note that $Q^{(i)} = y_1^iR_1[y_1,y_2] \cap S$. Hence, $R/Q^{(i)} = k[x_iy_1', x_iy_2: i = 1, 2, 3] \subset S_1[y_1, y_2]/(y_1^i)$, where y_1' is the image of y_1 modulo (y_1^i) . We shall show that x_1y_2, x_2y_2 is a regular sequence in $R/Q^{(i)}$, which will complete the proof. Suppose that $fx_2y_2 = gx_1y_2$. Then $fx_2 = gx_1$, and f, g break into bihomogeneous pieces of the same bidegree in y_1', y_2 for which this is true, and we might as well assume that $f = f_0y_1'^py_2^q$ and $g = g_0y_1'^py_2^q$, where $f_0, g_0 \in R_1$. Then $f_0x_2 = g_0x_1$ in R_1 , and since x_1, x_2 is an R_1-sequence, the result follows easily.

BIBLIOGRAPHY

1. H. Bass, On the ubiquity of Gorenstein rings, Math. Z. 82 (1963), 8-28.

2. D. Buchsbaum and D. Eisenbud, What makes a complex exact?, to appear in J. of Alg.

3. ________, Remarks on ideals and resolutions, preprint.

4. W. L. Chow, On unmixedness theorem, Amer. J. Math. 86 (1964), 799-822.

5. D. Ferrand and M. Raynaud, Fibres formelles d'un anneau local Noethérien, Annales Sci. de l'École Normale Supérieure 3 (1970), 295-312.

6. A. Grothendieck (with J. Dieudonné), Éléments de géométrie algébrique, IV. (Seconde partie), Publications mathématiques de l'I. H. E. S. n° 24, Paris, 1965.

7. A. Grothendieck (notes by R. Hartshorne), Local cohomology, Springer-Verlag Lecture Notes in Mathematics No. 41, 1967.

8. M. Hochster, Grade-sensitive modules and perfect modules, preprint.

9. ________, Contracted ideals from integral extensions, preprint.

10. I. Kaplansky, Commutative rings, Allyn and Bacon, Boston, 1971.

11. ________, Topics in commutative ring theory, I. and III., duplicated notes.

12. G. Levin and W. Vasconcelos, Homological dimensions and Macaulay rings, Pacific J. Math. 25 (1968), 315-323.

13. M. Nagata, Local rings, Interscience Tracts 13, New York, 1962.

14. C. Peskine and L. Szpiro, Dimension projective finie et cohomologie locale, Thesis (Orsay, Serie A, N° d'Ordre 781), to appear in Publ. I. H. E. S.

15. D. Rees, The grade of an ideal or module, Proc. of the Cambridge Philosophical Society 53 (1957), 28-42.

16. J. P. Serre, Algèbre locale. Multiplicités. Springer-Verlag Lecture Notes in Mathematics No. 11, 1965.

17. R. Y. Sharp, Application of dualizing complexes to finitely generated modules of finite injective dimension, preprint.

Comments and Corrections

1. There is a serious gap in the proof of the main theorem in [9] (see the first paragraph after Question 4.1). However, the result still holds if R contains a field, and the general case is a consequence of a weak form of Conjecture (E) . This will be shown in a revised version of [9].

2. It can be shown in the 2-dimensional case that the classes $[R_n/(s_1,s_2)]$, where s_1,s_2 is an R_n-sequence, generate $\lim_{\to n} \underline{G}(R_n)$, so that we need only divide out by these instead of all g-perfect modules to get a version of Conjecture (F) to hold.

COMMUTATIVE RINGS

Irving Kaplansky

1. INTRODUCTION

I have chosen to speak on the subject of commutative Noetherian rings, a topic which has fascinated me for years, and one which has grown to maturity as one of the basic "structures" of mathematics. I would like to take the opportunity to survey it in historical perspective. Let me hasten to observe that Bourbaki's seven chapters [6] are accompanied by a historical note up to his usual erudite standard. I will therefore touch comparatively lightly on the classical background, and spend more time on the recent history, which N. Bourbaki's authoritative pen has yet to record in detail.

The subject seems to me to divide into five well defined eras, or rather a "prehistoric" era and then four periods, each signalled by the appearance of a very important paper.

2. THE PREHISTORIC ERA

The field begins, as so much of 19th century mathematics did, with the *Disquisitiones*. Gauss's theory of binary quadratic forms amounted to a virtually complete account of the integers of a quadratic field. At the hands of Kummer, Dedekind, and Kronecker, the ring of integers of any algebraic number field became well established, and so did the parallel theory of algebraic functions of one variable. This is the "one-dimensional" case of our subject. For the n-dimensional case we have to look to the stream of work on algebraic geometry during the 19th century. A. Weil has aptly written, in the introduction to [40], that the algebraic geometers "rested on a most painstaking study of numerous special examples, from which they gained an insight not always found among modern exponents of the axiomatic creed".

3. THE FIRST PERIOD

The 19th century began with a monumental piece of work; it also ended with one.

Hilbert's memoir [14] made three extremely important contributions.

(1) He proved that if K is a field and $R=K[x_1,\ldots,x_n]$ the ring of polynomials in n variables, then every ideal of R is finitely generated.

Later this was recast and slightly generalized to the form we now call the *Hilbert basis theorem:* if in a commutative ring R every ideal is finitely generated, then the same is true in $R[x]$. Hilbert's decisive theorem united at one stroke a large number of finiteness observations that had previously been made case by case.

(2) If I is an ideal in $K[x_1,\dots,x_n]$ there is therefore a finite set of generators for I, say $f_1,\dots,f_r$. That is to say, any polynomial in I is expressible as a linear combination of $f_1,\dots,f_r$ with coefficients that are again polynomials. This expression will not, however, ordinarily be unique. The lack of uniqueness is measured by relations

$$u_1f_1+ \dots + u_rf_r = 0$$

called "syzygies" by the algebraic geometers. For deeper insight he next investigated the r-ples $u_1,\dots,u_r$. He recognized that he was now dealing with a module rather than an ideal, but showed it was again finitely generated and proceeded to the "second syzygies". Theorem: *after n steps you reach 0, or to state it otherwise, after n-1 steps a free module is reached.* This tour de force was a farsighted anticipation of homological algebra.

(3) Given an ideal I in $K[x_1,\dots,x_n]$ (actually Hilbert dealt only with homogeneous ideals, i.e., ideals which can be generated by forms) Hilbert studied the asymptotic number of linearly independent forms of a given degree contained in I, and was led to a certain *characteristic polynomial*. Subsequently this turned out to be a key tool for the modern rigorous treatment of intersection multiplicities of algebraic varieties.

Three years later, Hilbert [15] proved another fundamental theorem: *the Nullstellensatz*. I would like to try to interest the younger generation of functional analysts, so let me begin by recalling that if X is a compact Hausdorff space and $C(X)$ the ring of continuous complex functions on X, then every maximal ideal in $C(X)$ consists of all the functions vanishing at some point. Analogously, let $R=K[x_1,\dots,x_n]$ be the ring of polynomials in n variables over the algebraically closed field K, and think of R as consisting of functions on the space of n-tuples over K. Theorem: *every maximal ideal in R is the set of all polynomials vanishing at some point.*
The polynomial theorem is a good deal harder than the continuous function theorem. But here is a curious remark: the Nullstellensatz is easy if K is uncountable. (I say "curious" because, among other things, in Hilbert's

day one did not specify K, it being tacitly assumed that K was the field of complex numbers.) The proof is contained in [19] and [20], but let me briefly repeat it. Given a maximal ideal M in R, our problem can be recast to the statement that R/M coincides with K. Now if the field R/M is larger than K it must (since K is algebraically closed) contain a transcendental element u. It is a fact (and a nice exercise in linear algebra) that the elements $1/(u-a)$, a ranging over K, are linearly independent. This is a contradiction if K is uncountable, for R is a vector space of countable dimension over K.

The Nullstellensatz sets up a one-to-one correspondence between maximal ideals and points. From this one can pass to a one-to-one correspondence between prime ideals and irreducible algebraic varieties. This is of course a vital step in building the bridge between algebra and geometry.

The needs of both algebra and geometry require us to study all ideals, not just the prime ideals. The next step was taken by Lasker [21] when he proved that any ideal admits an expression as an intersection of primary ideals. This theorem is as good an analogue as we can expect of the representation of an integer as a product of prime powers. Important additions were made by Macaulay in a number of papers leading up to his Cambridge tract [23]. In many respects Macaulay was far ahead of his time, and some aspects of his work won full appreciation only recently. One cannot fail to be impressed with the fact that Krull himself [17, p.68] found Macaulay's work to be "teilweise recht mühsam".

4. THE SECOND PERIOD

The second era was inaugurated by Emmy Noether's paper [28]. The importance of this paper is so great that it is surely not much of an exaggeration to call her the mother of modern algebra.

After a leisurely description of the background she takes as a sole axiom the conclusion of Hilbert's basis theorem: *that every ideal is finitely generated, or equivalently the ascending chain condition on ideals*. A very simple, general argument shows that any ideal is the intersection of a finite number of irreducible ideals (i.e. ideals not expressible as an intersection of two larger ones). And then comes the charming argument that ushered in the new era: a short proof that any irreducible ideal is primary.

It is entirely appropriate that rings with the ascending chain condition were later called "Noetherian".

Let me take the space to record the proof. I do it with the harmless normalization that the irreducible ideal is 0. So: we have a Noetherian ring in which 0 is irreducible and have to prove that every zero-divisor is nilpotent (for this is what it means to say that 0 is a primary ideal). Let $ab=0$ with $b\neq 0$. Let I_n be the annihilator of a^n. The ideals $\{I_n\}$ ascend, and therefore ultimately become stable, say at I_k. We claim $I_k \cap Ra^k=0$. For suppose $x=ya^k$ lies in the intersection. Then $0=xa^k=ya^{2k}$, $y \in I_{2k}$. But $I_{2k}=I_k$, so $ya^k=0$, $x=0$. By the irreducibility of 0, either I_k or Ra^k is 0. But I_k contains b. Hence $Ra^k=0$ and a is nilpotent as required.

In a subsequent paper [29], Emmy Noether gave her celebrated axiomatic treatment of rings with the properties found in algebraic number theory (subsequently called Dedekind rings). Before long, van der Waerden's book [37] taught all these things to a whole generation of algebraists.

I insert at this point a parenthetical and controversial note. In 1927 Artin [1] introduced what one might call the dual class of rings; those with the *descending* chain condition. For forty years these *Artinian* rings have been regarded as the natural way to generalize the subject of finite-dimensional algebras to a piece of ring theory. But I am beginning to doubt it. Natural examples of Artinian rings are not common. Infinite-dimensional division rings, perhaps the most tangible of the Artinian rings which are not finite-dimensional algebras, are hard to work with and remain largely mysterious. I argue that, with commutative Noetherian rings and finite-dimensional algebras over fields substantially under control, the next natural step is the class of (possibly non-commutative) rings which are finitely generated modules over commutative Noetherian rings. Among other things, we need a name for this class of rings!

5. THE THIRD PERIOD

In the early 1920's a strong new voice arose to carry on the tradition. Right up to the present day Krull has investigated every aspect of commutative rings with penetration and discernment. His Ergebnisse monograph of 1935 [17] gave a broad view of the subject as he saw it at the time and makes stimulating reading today.

The third era was inaugurated by the paper [16] in which Krull proved the *principal ideal theorem.* (Warning to the group-theorists: this has nothing to do with the principal ideal theorem of group theory. To make matters worse Krull gives this name to two theorems: see pages 25 and 37 of [17]. The one on page 25 is a comparative triviality.)

As Northcott justly remarks [30, pp. 105-106], until the principal ideal theorem arrived, Noetherian rings were but a pale shadow of polynomial rings. To this day it remains a kind of *pons asinorum* of the subject. It is a testimonial to Bourbaki's thoroughness that his seven chapters have not yet reached the principal ideal theorem.

To state the theorem, I introduce the *rank* of a prime ideal, saying that P has rank n if there exists a chain $P=P_0 \supset P_1 \ldots \supset P_n$ of prime ideals of length n, but no longer chain. Theorem: *in a Noetherian ring, a prime ideal which is minimal over a principal ideal has rank* ≤ 1. From this one passes relatively easily to the king-size principal ideal theorem; a prime ideal minimal over an ideal generated by n elements has rank $\leq n$.

Krull was of course inspired by the known facts about polynomial rings and their homomorphic images. Briefly put, the principal ideal theorem is in that context a consequence of classical properties of the dimension of an algebraic variety and of the dimension of an intersection.

I note that Rees gave a sparkling new proof of the principal ideal theorem in [31], a paper that also marked the appearance of the Artin-Rees lemma.

The principal ideal theorem yields the information that the prime ideals in a Noetherian ring satisfy the descending chain condition in a strengthened form, to wit with a uniform bound on chain descending from a fixed prime. Of course, by axiom we have the ascending chain on all ideals and a fortiori on prime ideals, but perversely enough there need not be a uniform bound on chains ascending from a fixed prime.

I see the next landmark as Krull's 1938 paper [18] on local rings. The technique of localization was getting established, although it was not until Uzkov's paper [36] in 1946 that it arrived in complete generality. In particular, localization at a maximal ideal reduces many problems to the case of a single maximal ideal. Krull called such a ring a "Stellenring", but Chevalley's renaming to "local ring" has stuck. It is to be noted that "local" is here meant to incorporate Noetherian. It is reasonable in mathematics that the most important concept should have the short name. But two difficulties have arisen: the convention is not universally accepted, and the dreadful circumlocution "not necessarily Noetherian local ring" can be found in the literature. I suggest that the designation "quasi-local" for a commutative ring with one maximal ideal would be a reasonable choice on which to agree.

The two major accomplishments in [18] are the introduction of the technique of completing a local ring, and the recognition of a certain special class of local rings. Recall that if the maximal ideal M of a local ring R can be generated by n elements, then n is the maximal length of a chain of prime ideals in R. Krull called R a "p-Reihenring" if this maximum is attained. Another name was doubtless in order and Chevalley's [7] "regular" is now universally accepted. It was perhaps not the happiest choice since, at the latest count, "regular" has 2,947 meanings in mathematics; and, worst of all, there was already at that time a definition due to von Neumann [39] of an entirely different concept of regularity for rings.

REMARK. This account is over-simplified: Krull and Chevalley used alternative definitions for regularity, and furthermore Chevalley restricted his local rings somewhat.

It turned out that regularity corresponded to a very classical concept indeed, that of a nonsingular point on a variety. This was fully set forth by Zariski [41].

Two questions were promptly raised in Krull's original paper.

(1) If R is regular, and P is a prime ideal in R, is R_P (the localization of R relative to P) regular?

(2) Is a regular local ring a unique factorization domain?

For the regular local rings that arise in algebraic geometry, the geometrical interpretation settles (1) in the affirmative rather easily. With additional effort, Zariski [41, p. 22] also succeeded in answering the second affirmatively for geometric regular local rings. But the general case, for both problems, remained open.

The late I.S. Cohen's thesis [9] provided a kind of climax for this sequence of developments. He gave structure theorems for complete local rings, especially regular complete local rings, which came close to being decisive. The subject now had a method for launching a systematic attack on a proposed problem: localize at a general maximal ideal, complete, use Cohen's structure theory, then (hopefully) work your way back to the original ring. The method works with gratifying frequency, and every young ringtheorist should learn how to use it. But it does not always work, e.g. the two problems mentioned above continued stubbornly to hold out.

A pattern began to appear repeatedly. A certain theorem would get proved for a special class of Noetherian rings, usually something like the geometric case. Workers in the field relentlessly accepted the challenge to remove the restriction. This was partly for the usual reason: the elegance and clarity of a good generalization. But partly it was in answer to current and anticipated needs; for instance, the subject of diophantine geometry calls for a study of polynomial rings over the integers and their homomorphic images.

I will give an explicit instance which is perhaps a little more interesting since in this case the theorem did not generalize. Let R be a geometric local domain. Chevalley [8] proved that the completion R^* has no nilpotent elements. This is not true for a general local domain, not even if R is integrally closed [26]. But Rees, in a splendid paper [33], pinpointed the *exact* condition needed: for every ring T between R and its quotient field which is finitely generated over R, we must have that the integral closure of T is a finitely generated T-module. Of course it is standard that this condition is fulfilled in the geometric case.

Incidentally, there is a companion theorem due to Zariski [42]: *if R is an integrally closed geometric local domain then the completion R^* is again a domain.* Nagata's example in [26] again rules out a complete generalization. It would be nice to have a companion to Rees's theorem, showing exactly what is needed for Zariski's theorem to hold.

6. THE FOURTH PERIOD

We are living in the fourth era right now, and its advent was heralded by the arrival of homological algebra.

Let us recall as much as we need. Given an R-module A map a projective module (i.e. a direct summand of a free module) onto A, getting a kernel K_1, then map a projective module onto K_1 getting a kernel K_2, etc. If the first of the K's to be projective is K_n we say that the homological dimension of A is n, and write $d(A) = n$; the value of n does not depend on the way the process was carried out. If no K_n is projective we write $d(A) = \infty$. Note the resemblance to Hilbert's chains of syzygies.

The big theorem was proved by Auslander, Buchsbaum and Serre. (The Auslander-Buchsbaum portion was announced in [2], with full details in [3]; Serre finished the job in [34].) Let R be local with maximal ideal M. Then: R is regular if and only if $d(M) < \infty$. For reference a little later we note that it follows further that when R is regular, $d(A) < \infty$ for *every* R-module A.

Now it turns out, for the technical reason that homological dimension behaves very well under localization, that this characterization of regularity is easily proved to be inherited by R_P. Thus the first of Krull's problems was solved. This resounding triumph of the new homological method marked a turning point of the subject of commutative Noetherian rings.

Krull's problem on unique factorization had to wait four more years till it too yielded to Auslander and Buchsbaum [4]. Let me describe the form this solution took at the hands of MacRae. One translates the problem into the following: if I is an ideal generated by two elements, prove $d(I) \leq 1$. (Explanation: map a free module onto the two generators of I, getting a kernel K. Since $d(I) \leq 1$, K is projective; and since R is local, K is free, necessarily on one generator. This gives us the greatest common divisor needed to prove unique factorization.) Now suppose one could prove $d(I) = 0, 1$, or ∞. The regularity hypothesis would rule out ∞, and all would be done. MacRae has proved the desired theorem, which we restate: *if I is any ideal generated by two elements in a commutative Noetherian ring, then* $d(I)=0$, 1, *or* ∞. He did it first for domains in [24]; in [25], with the aid of suggestions from M. Auslander and D. Mumford, he completed the proof.

Lamentably (or not, depending on your point of view) there is no theorem of this kind for ideals generated by three or more elements.

Many of the subsequent developments might be described as "introspective" that is, the homological tools were sharpened for their own sake. But I have no doubt that more applications await future discovery. Somewhat arbitrarily, I have selected five of these developments for a brief discussion.

1. Tate [35] gave a quantitative sharpening of the Auslander-Buchsbaum-Serre theorem (ABS). With M the maximal ideal of a local ring R, let B_k (the kth "Betti number") denote the number of elements needed to generate the $(k-1)$st kernel, when successive kernels are formed as described above. In particular, $B_1 (=n$, say) is the minimal number of generators of M. ABS says that if some $B_k=0$ then R is regular. Now one knows that B_k is at least as large as the binomial coefficient ${}_nC_k$. Tate proves: if $B_k={}_nC_k$ for a single $k \geq 2$ then R is regular, and he and others have sharpened this considerably. A challenging problem is to settle whether the Poincaré series $\sum B_k Z^k$ is a rational function of Z; this is known in a number of special cases.

Tate's technique was to form an algebra resolution of M, and he asked if it could be chosen to be minimal in an appropriate sense. Just recently, Gulliksen [13] answered this in the affirmative.

2. There is a dual theory of homological dimension, based on injective modules. Bass [5] studied the class of Noetherian rings R with the

property that the module R itself has finite injective dimension. (One does not do this on the projective side of the ledger, for of course R has projective dimension 0.) It turned out that the condition coincided with one that Gorenstein [11] had introduced for algebraic geometric purposes, and so these rings are called Gorenstein rings. Bass surveyed a surprisingly large number of connections with other topics in Noetherian rings.

3. Like any good theorem, ABS deserves generalizations. The following theorem due to Levin is published in his joint paper with Vasconcelos [22]. Suppose for the maximal ideal M of a local ring R we have $d(M^n)<\infty$ for some power M^n. We cannot conclude that R is regular, for M^n can equal 0 in a nonregular R. But: this is the only exception.

4. The theorem proved by Vasconcelos in [38] concerns R-sequences. An R-sequence is, so to speak, a sequence of nonzero-divisors. More exactly, the sequence $a_1,\ldots,a_n$ in R is an R-sequence if a_1 is a nonzero-divisor in R, a_2 maps into a nonzero-divisor in $R/(a_1)$, a_3 maps into a nonzero-divisor in $R/(a_1,a_2)$, etc. R-sequences made their appearance in a very natural way in the initial stages of the work of Auslander and Buchsbaum [3] and of Rees [32]. They deserve an extensive report of their own, but I shall merely emphatically say that R-sequences are here to stay.

I quote still another characterization of the regularity of a local ring R with maximal ideal M; this time the condition is that M should be capable of being generated by an R-sequence. Now we are ready for Vasconcelos's theorem: *for an ideal I in a local ring, I can be generated by an R-sequence if and only if $d(I)<\infty$ and I/I^2 is a free (R/I)-module.* The application to ABS is immediate, for when $I=M$, M/M^2 is a vector space over the field R/M and is certainly free.

5. This fifth remark is an open question which has been something of a challenge to the Chicago school: if R is local, A an R-module with $d(A)<\infty$, and x is a zero-divisor in R, does it follow that x is a zero-divisor on A?

7. CONCLUSION

There are many more topics that should be surveyed in a full report, and I hope an expert will write one soon. As examples I mention Zariski's main theorem on algebraic correspondences, Nagata's solution of Hilbert's fourteenth problem, Nagata's ingenious counter-examples to a flock of classical problems, the resolution of singularities and Hironaka's proof for characteristic 0, the formula for intersection multiplicities as an alternating sum of dimensions of Tor's, Lichtenbaum's solution of the problem of the rigidity of Tor, Samuel's investigations on unique factorization domains and his notable examples of failure of power series rings, and finally the generalization by the Grothendieck school of everything to schemes, as required by the new foundations of algebraic geometry (a scheme being, crudely speaking, a number of rings appropriately pasted together). But I hope I have given a useful picture of the field of commutative Noetherian rings and its status in the contemporary mathematical scene.

BIBLIOGRAPHY

The bibliography is confined to items to which there is an actual reference, except that I have added texts that a reader may wish to consult: [10], [12], [27], and [43].

1. E. Artin, *Zur Theorie der hypercomplexen Zahlen*, Abh. Math. Sem. Univ. Hamburg 5 (1927), 251-260.
2. M. Auslander and D.A. Buchsbaum, *Homological dimension in Noetherian rings*, Proc. Nat. Acad. Sci. USA 42 (1956), 36-38.
3. ________, *Homological dimension in local rings*, Trans. Amer. Math. Soc. 85 (1957), 390-405.
4. ________, *Unique factorization in regular local rings*, Proc. Nat. Acad. Sci. USA 45 (1959), 733-734.
5. H. Bass, *On the ubiquity of Gorenstein rings*, Math. Z. 82 (1963), 8-28.
6. N. Bourbaki, *Algèbre commutative*, Ch. I-VII, Hermann, Paris, 1961-1965.
7. C. Chevalley, *On the theory of local rings*, Ann. of Math. 44 (1943), 690-708.
8. ________, *Intersections of algebraic and algebroid varieties*, Trans. Amer. Math. Soc. 57 (1945), 1-85.
9. I.S. Cohen, *On the structure and ideal theory of complete local rings*, Trans. Amer. Math. Soc. 59 (1946), 54-106.

10. J. Dieudonné, *Topics in Local Algebra*, Notre Dame Mathematical Lectures no. 10, Notre Dame Univ. Press, Notre Dame, Indiana, 1967.

11. D. Gorenstein, *An arithmetic theory of adjoint plane curves*, Trans. Amer. Math. Soc. 72 (1952), 414-436.

12. A. Grothendieck (avec J. Dieudonné), *Eléments de Géométrie Algébrique*, Ch. IV, Inst. Hautes Etudes Sci. Publ. Math., 20, 24, 28 (1964-5-6).

13. T.H. Gulliksen, *A proof of the existence of minimal R-algebra resolutions*, Acta Math. 120 (1968), 53-58.

14. D. Hilbert, *Über die Theorie der algebraischen Formen*, Math. Ann. 36 (1890), 473-534.

15. ________, *Über die vollen Invariantensysteme*, Math. Ann. 42 (1893), 313-373.

16. W. Krull, *Primidealketten in allgemeinen Ringbereichen*, S.-B. Heidelberger Akad. Wiss. Math.-Natur. Kl. (1928), 7.

17. ________, *Idealtheorie, Ergebnisse der Mathematik und ihrer Grenzgebiete*, Springer-Verlag, Berlin, 1935.

18. ________, *Dimensionstheorie in Stellenringen*, J. Reine Angew. Math. 179 (1938), 204-226.

19. ________, *Jacobsonsche Ringe, Hilbertscher Nullstellensatz, Dimensionstheorie*, Math. Z. 54 (1951), 354-387.

20. S. Lang, *Hilbert's Nullstellensatz in infinite-dimensional space*, Proc. Amer. Math. Soc. 3 (1952), 407-410.

21. E. Lasker, *Zur Theorie der Moduln und Ideale*, Math. Ann. 60 (1905), 20-116.

22. G. Levin and W. Vasconcelos, *Homological dimensions and Macaulay rings*, Pacific J. Math. 25 (1968), 315-323.

23. F.S. Macaulay, *Algebraic theory of |modular| systems*, Cambridge Tracts no. 19, Cambridge, 1916.

24. R. MacRae, *On the homological dimension of certain ideals*, Proc. Amer. Math. Soc. 14 (1963), 746-750.

25. ________, *On an application of the Fitting invariants*, J. Algebra 2 (1965), 153-169.

26. M. Nagata, *An example of normal local ring which is analytically ramified*, Nagoya Math. J. 9 (1955), 111-113.

27. ________, *Local Rings*, *Interscience*, New York, 1962.

28. E. Noether, *Idealtheorie in Ringbereichen*, Math. Ann. 83 (1921), 24-66.

29. __________, *Abstrakter Aufbau der Idealtheorie in algebraischen Zahl- und Funktionenkörpern*, Math. Ann. 96 (1927), 26-61.

30. D.G. Northcott, *Ideal Theory*, Cambridge Univ. Press, New York 1953.

31. D. Rees, *Two classical theorems of ideal theory*, Proc. Cambridge Philos. Soc. 52 (1956), 12-16.

32. __________, *A theorem of homological algebra*, Proc. Cambridge Philos. Soc. 52 (1956), 605-610.

33. __________, *A note on analytically unramified local rings*, J. London Math. Soc. 36 (1961), 24-28.

34. J.-P. Serre, *Sur la dimension homologique des anneaux et des modules Noethériens*, Proc. Internat. Sympos. Algebraic Number Theory, Tokyo, 1955, pp. 175-189.

35. J.F. Tate, *Homology of Noetherian rings and local rings*, Illinois J. Math. 1 (1957), 14-27.

36. A.I. Uzkov, *On rings of quotients of commutative rings*, Mat. Sbornik N.S. 13 (1943), 71-78.

37. B.L. van der Waerden, *Moderne Algebra*, vol. II, Springer-Verlag Berlin, 1931. (Also later editions and English translation.)

38. W. Vasconcelos, *Ideals generated by R-sequences*, J. Algebra 6 (1967), 309-316.

39. J. von Neumann, *On regular rings*, Proc. Nat. Acad. Sci. USA 22 (1936), 707-713.

40. A. Weil, *Foundations of Algebraic Geometry*, Amer. Math. Soc. Colloq. Publ. vol. 29, Amer. Math. Soc. Providence, R.I., 1946, Rev. ed. 1962.

41. O. Zariski, *The concept of a simple point of an abstract algebraic variety*, Trans. Amer. Math. Soc. 62 (1947), 1-52.

42. ________, *Analytical irreducibility of normal varieties*, Ann. of Math. 49 (1948), 352-361.

43. O. Zariski and P. Samuel, *Commutative Algebra*, Van Nostrand, Princeton, N.J. vol. I, 1958, vol., II 1962.

University of Chicago
Chicago, Illinois

Addendum (July, 1972)

This lecture was originally presented in June, 1968 as the first Jeffery-Williams lecture of the Canadian Mathematical Congress. It was published in 1972 by the Canadian Mathematical Congress, along with other lectures of this series. Thanks are expressed to the Congress, and to John J. Mc Namee, their Executive Director, for permission to reprint it here.

I shall take the opportunity to update the lecture a little, beginning with the fact that Krull died in 1971. His monograph [17] appeared in a second edition, revised and expanded, in 1968.

In EGA, the Grothendieck-Dieudonné Elements [12], a fourth part of Chapter IV appeared as no. 32 in the IHES series. Springer's reprinting of EGA, in book form and slightly revised, has begun with the appearance in 1971 of Chapters 0 and 1. It should be added that §6 of Chapter 0 consists of significant "Compléments" concerning commutative algebra.

Here is an indication of the order of magnitude faced by a person planning to update himself in commutative ring theory since 1968. With the Jan. 1968 issue, Mathematical Reviews began to have a section titled "Commutative associative rings and algebras". Since then, to the middle of 1972, approximately 750 items have appeared under this heading. I shall content myself with listing the books that have appeared (in addition to my Commutative Rings, Allyn and Bacon, 1970). In this list I am interpreting commutative ring theory rather narrowly.

M.F. Atiyah and I.G. MacDonald, Introduction to Commutative Algebra, Addison-Wesley, 1969.

R.W. Gilmer, Multiplicative Ideal Theory, Queens Papers on Pure and Applied Mathematics, no.12, 1968. In revised and expanded form, this will shortly be published by Dekker.

M.D. Larsen and P.J. McCarthy, Multiplicative Theory of Ideals, Academic Press, 1971.

H. Matsumura, Commutative Algebra, Benjamin, 1970.

D.G. Northcott, Lessons on Rings, Modules, and Multiplicities, Cambridge, 1968.

J.T. Knight, Commutative Algebra, Lon. Math. Soc. Lecture Note Series No. 5, Cambridge, 1971.

ON EUCLIDEAN RINGS OF ALGEBRAIC FUNCTIONS

by R. E. MacRae*

University of Colorado, Boulder, Colorado

ABSTRACT. A large class of rings of algebraic functions are shown to be principal ideal domains but not Euclidean with respect to any possible algorithm.

1. Introduction. A Euclidean ring for our purposes will be an integral domain together with a natural number valued function φ defined on the non-zero elements of the ring such that for all α and β in the ring with $\beta \neq 0$ there exist γ and δ in the ring such that $\alpha = \beta\gamma + \delta$ and either $\delta = 0$ or $\varphi(\delta) \lneq \varphi(\beta)$. It is of course well known that all such rings are principal ideal domains. A rather curious fact, however, is that, while principal ideal domains arise with great frequency in all sorts of arithmetic and geometric contexts, relatively few of these are known to be Euclidean. It is the purpose of this paper to show that in the case of the rings of algebraic functions defined on an algebraic curve they are almost never Euclidean. To be more precise, let k be an infinite field and K a function field in one variable whose exact constant field is k. Let $X^{K/k}$ be the Riemann surface for K over k and let U be an open subset of $X^{K/k}$. It is well known that the ring $\mathcal{O}(U) = \{\alpha \in K \mid \alpha$ has no poles on $U\}$ is a Dedekind domain. In fact U need not even be open. See, for example, [2]. When k is a finitely generated extension field of a finite field or of the rational numbers then the Mordell-Weil Theorem is valid and it is easily seen that every open set U contains an open set V such that $\mathcal{O}(V)$ is a principal ideal domain. Consequently, the cases in which $\mathcal{O}(U)$ is principal are very numerous. We will, however, prove the following results.

Theorem A: If only finitely many of the points on the Riemann surface for K over k are k-rational then $\mathcal{O}(U)$ is not Euclidean for any proper open set U.

It should be remarked that when k is a number field and the genus of K over k is at least two then all known examples satisfy the hypothesis of Theorem A. The Mordell conjecture in fact asserts that all curves of genus at least two over a number field k have only finitely many k-rational points and so Theorem A applies.

---

*This work was in part supported by a National Science Foundation Grant.

Theorem B: Let K be a function field in one variable whose exact constant field is k and whose genus is at least two and an invariant under extensions of k. Moreover, assume that k is a function field in one variable over the finite field $k_0 = GF(q)$ whose genus is strictly less than that of K. Then, if K possesses no non-trivial automorphisms over any extension of k, it follows that $\mathcal{O}(U)$ is not Euclidean for any proper open set U on the Riemann surface for K over k.

It should be remarked that this result includes a very large number of cases since the assumption of the triviality of the automorphism group appears to hold for most curves of large genus.

I am indebted to P. Samuel for a number of helpful conversations connected with these problems.

2. Main Results. We make use of the following definition and results from [3].

Definition 2.1: An integral domain $\mathcal{O}$ is called Euclidean if there is a function $\varphi: (\mathcal{O} - \{0\}) \to N$ such that for all α, β in $\mathcal{O}$ with $\beta \neq 0$ there exist γ and δ in $\mathcal{O}$ such that $\alpha = \beta\gamma + \delta$ and $\delta = 0$ or $\varphi(\delta) \lneq \varphi(\beta)$.

Proposition 2.2: If M is a multiplicatively closed subset of the Euclidean ring $\mathcal{O}$ then the localized ring $\mathcal{O}_M$ is also Euclidean.

See [3, Prop. 7] for a proof.

Theorem 2.3: Let k be an infinite field and K a finite extension field of k. If G is a finitely generated subgroup of K^* such that $K^* = k^*G$ then $K = k$.

See [3, Prop. 18] for a proof.

We can now give a proof of our first result.

Theorem A: Let K be a function field in one variable over the infinite field k. If the Riemann surface for K contains only finitely many k-rational points then $\mathcal{O}(U)$ is not Euclidean for any proper open set U on the Riemann surface.

Proof: Suppose some $\mathcal{O}(U)$ is Euclidean. Let V be an open subset of U that excludes the finitely many k-rational points. Then $\mathcal{O}(V)$ is also Euclidean by Proposition 2.2 since $\mathcal{O}(V)$ is a localization of $\mathcal{O}(U)$. Moreover the group of units $\mathcal{O}(V)^*$ of $\mathcal{O}(V)$ is finitely generated since the complement of V in the Riemann surface is a finite set. Now since $\mathcal{O}(V)$ is Euclidean the first of the Motzkin conditions asserts that there is an element π of $\mathcal{O}(V)$ such that the natural map of $\mathcal{O}(V)^*$ to $(\mathcal{O}(V)/\pi\mathcal{O}(V))^*$ is onto. See [3]. But $\mathcal{O}(V)/\pi\mathcal{O}(V)$ is a field extension of finite degree over k and hence the Theorem 2.3 of Samuel applies and shows that $\mathcal{O}(V)/\pi\mathcal{O}(V) = k$. This is, however, impossible since none of the k-rational points of K is in V.

Our next result is somewhat deeper since it makes essential use of not only Theorem A but of the so-called Mordell conjecture for curves defined over curves. See, for example, [4].

Theorem B: Let K be a function field in one variable whose exact constant field is k and whose genus is at least two and an invariant under extensions of k. Moreover, assume that k is a function field in one variable over the finite field $k_0 = GF(q)$ whose genus is strictly less than that of K. Then, if K possesses no non-trivial automorphisms over any extension of k, it follows that $\mathcal{O}(U)$ is not Euclidean for any proper open set U on the Riemann surface for K over k.

Proof: We claim that K has only finitely many k-rational points. If this be proved, then Theorem A applies and we are done. Suppose not. Let $\bar{k}_0$ be the algebraic closure of k_0 and set $\bar{k} = \bar{k}_0 \otimes_{k_0} k$ and $\bar{K} = \bar{k} \otimes_k K$. It is clear that $\bar{k}$ and $\bar{K}$ are fields since $\bar{k}_0$ and $\bar{k}$ are algebraic extensions of k_0 and k, respectively. Moreover $\bar{K}$ has infinitely many $\bar{k}$-rational points since K was assumed to have infinitely many k-rational points. Since the genus of $\bar{K}$ over $\bar{k}$ is at least two, the theorem of Grauert applies [4, Theorem 1, pg. 107]. That is to say, there is a finite algebraic extension $\bar{k}'$ of $\bar{k}$ such that $\bar{K}' = \bar{k}' \otimes_{\bar{k}} \bar{K}$ is the quotient field of the ring $\bar{k}' \otimes_{\bar{k}_0} \bar{k}''$ where $\bar{k}''$ is a function field in one variable over $\bar{k}_0$ and is a subfield of $\bar{K}'$. Now $\bar{k}''$ can certainly be described by a curve defined over some finite field since $\bar{k}_0$ is the algebraic closure of a finite field. Hence the theorem of Grauert-Samuel applies [4, Theorem 2, pg. 109]. That is to say, $\bar{k}'$ can be picked to be a finite Galois extension of $\bar{k}$ and the Galois group of $\bar{k}'$ over $\bar{k}$ can be regarded as a subgroup of the automorphism group of $\bar{K}'$. The latter group is trivial, by hypothesis. Hence $\bar{k}' = \bar{k}$. Let, now, $\mathcal{O}_{\mathfrak{P}}$ be the valuation ring of some $\bar{k}$-rational point of $\bar{K} = \bar{K}'$. Two cases arise. We have either $\bar{k}'' \leq \mathcal{O}_{\mathfrak{P}}$ or not. Suppose that the first case holds. Then $\bar{k}''$ is a subfield of the residue class field $\bar{\mathcal{O}}_{\mathfrak{P}} = \bar{k}$. However, the genus of $\bar{k}''$ over $\bar{k}_0$ equals that of $\bar{K}$ over $\bar{k}$ while the genus of $\bar{k}$ over $\bar{k}_0$ is strictly less by hypothesis. Now let $\bar{k}''_s$ be the separable closure of $\bar{k}''$ in $\bar{k}$. Since $\bar{k}''$ and $\bar{k}$ are separably generated over the perfect field $\bar{k}_0$, $\bar{k}''_s = \bar{k}^{p^e}$ and thus the genus of $\bar{k}''_s$ over $\bar{k}_0$ equals that of $\bar{k}$ over $\bar{k}_0$. The relative genus formula [1, pg. 134] applies, therefore, to the pair of fields $\bar{k}''$ and $\bar{k}''_s$. This implies that the genus of $\bar{k}''$ is at most that of $\bar{k}''_s$ and we have a contradiction. Hence it is not possible that $\bar{k}''$ be contained in $\mathcal{O}_{\mathfrak{P}}$. Hence the $\bar{k}$-rational points of $\bar{K}$ arise entirely from $\bar{k}_0$-rational points of $\bar{k}''$. Now, by hypothesis, infinitely many of the $\bar{k}$-rational point of $\bar{K}$ are defined over k. Hence we can find a finite subfield of $\bar{k}_0$ over which these $\bar{k}$-rational points arise as $\bar{k}_0$-rational points of $\bar{k}''$. This is, however, impossible and we have a final contradiction.

References

1. M. Eichler, Introduction to the Theory of Algebraic Numbers and Functions, Academic Press, New York 1966.

2. R. E. MacRae, On unique factorization in certain rings of algebraic functions, J. of Alg., 17(1971), 243-261.

3. P. Samuel, About Euclidean rings, J. of Alg., 19(1971), 282-301.

4. P. Samuel, Lectures on Old and New Results on Algebraic Curves, Tata Institue, Bombay, 1966.

SUBFIELDS OF INDEX 2 OF ELLIPTIC FUNCTION FIELDS

by R.E. MacRae* and Pierre Samuel
University of Colorado

This is an account of some recent work done by us on elliptic function fields. The writer (P.S.), not being an expert on these questions, suspects that part of our results is already known. Our work is only at its preliminary stage, and many problems are still open.

By "curve" we mean a nonsingular projective algebraic curve. Points of curves have their coordinates in an algebraically closed field, as big as needed (we do not care about inflation and resources exhaustion, at least as far as transcendental elements are concerned!) If k is a subfield of this algebraically closed field over which the curve C is defined, C_k denotes the set of k-rational points of C. We exclude the case of characteristic 2, which has quite a different flavor. From time to time, characteristic 3 will also be excluded. In sections I to V, C is an elliptic curve defined over a field k (char $(k) \neq 2$) and C_k is non-empty.

§I - Introduction

We fix an origin on C, so that C is a group for the well known addition of points. Any subfield K of genus 0 of $k(C)$ is purely transcendental over k. We are interested in the subfields K of genus 0 and index 2 of $k(C)$ (i.e. $[k(C):K] = 2$). The following results are well known or easy (non-exclusive"or"):

(1.1) - K is the field of invariants of some involution $\tau \in \mathrm{Aut}_k(C)$, geometrically defined as follows: There is a unique point $A \in C_k$ such that, for all $M \in C$, we have $\tau(M) = -A - M$.

(1.2) - If we view C as a plane cubic curve $y^2 = x^3 + ax^2 + bx + c$, and if (u,v) are the coordinates of A, then $K = k(\frac{y-v}{x-u})$ (as usual, we take for origin on C its unique point at infinity).

*This author was, in part, supported by a grant from the National Science Foundation.

(1.3) - For any generating element t of K (i.e. $K = k(t)$), we may write $k(c) = k(s,t)$ with $s^2 = P(t)$, P being a <u>squarefree quartic polynomial</u> over k (a cubic polynomial is viewed here as a quartic polynomial with a root at infinity).

(1.4) - The roots of $P(t)$ correspond to the <u>branch points</u> of $k(C)$ over K, or (geometrically) to the 4 tangents drawn from A to the cubic curve C. The contact points M of these tangents are the solutions of the equation $\tau(M) = M$, i.e. $2M + A = 0$.

(1.5) - If the roots of $P(t)$ are written in a suitable order t_1, t_2, t_3, t_4, their <u>cross-ratio</u> (t_1,t_2,t_3,t_4) has a value r independent of K; at any rate, it takes one of the values $r, \frac{1}{r}, 1-r, 1-\frac{1}{r}, \frac{1}{1-r}, \frac{r}{r-1}$. This set of values characterizes the <u>isomorphism class</u> of C over the algebraic closure of k.

(1.6) - This set has actually 6 elements, except in the cases:

(H) It is $\{-1, 2, \frac{1}{2}\}$, in which case we say that C, or $P(t)$ is <u>harmonic</u>;

(EH) It is $\{-j, -j^2\}$ $(j^3 = 1)$ in which case we say that C or $P(t)$ are <u>equianharmonic</u> (case (EH) does not occur in characteristic 3).

Cases (E) and (EH) are the only ones in which C admits other <u>automorphisms</u> than the translations $(M \mapsto B + M)$ and the involutions $(M \mapsto B - M)$: if we call $\mathcal{T}$ the group of translations of C, then $\mathrm{Aut}(C)/\mathcal{T}$ is cyclic of order 4 (resp.6), provided the characteristic is $\neq 3$ (In characteristic 3 and case (H), $\mathrm{Aut}(C)/\mathcal{T}$ is a dihedral group of order 12). The cubic equation of C and a generating element of $\mathrm{Aut}(C)$ modulo $\mathcal{T}$ are as follows (char $\neq 3$):

$$\text{(H)} \quad y^2 = x^3 + bx\,, \quad (x,y) \mapsto (-x, iy) \quad (i^2 = -1)$$

$$\text{(EH)} \quad y^2 = x^3 + c\,, \quad (x,y) \mapsto (jx, -y) \quad (j^3 = 1)$$

If these automorphisms are defined over k (i.e. if $i \in k$, resp. $j \in k$), we say that C is k-<u>harmonic</u> (resp. k-<u>equianharmonic</u>).

§II - <u>Conjugate subfields of</u> $k(C)$

We say that two subfields of genus 0 and index 2, K and K' of $k(C)$ are conjugate if there is a $\sigma \in \mathrm{Aut}_k(C)$ such that $\sigma(K) = K'$. If τ, τ' are the corresponding involutions, this means

(2.1) $$\tau' = \sigma\tau\sigma^{-1}$$

Let A, A' be the points of C_k such that $\tau(M) = -A-M$, $\tau'(M) = -A'-M$. Then:

Proposition 2.2 - If $\mathrm{Aut}_k(C)$ contains only translations and involutions, K and K' are conjugate iff $A-A' \in 2C_k$. The conjugacy classes of subfields correspond to the elements of $C_k/2C_k$.

In fact, if σ is the translation $M \mapsto B+M$ (resp. the involution $M \mapsto B-M$), (1) means $A-A' = 2B$ (resp. $A+A'+2B = 0$, i.e. $A-A' = -2(B+A')$). Q.E.D.

When C is k-harmonic or k-equianharmonic, things are a little more complicated. If we call m a generating element of $\mathrm{Aut}_k(C)$ modulo the translations, then a simple computation shows that the conjugacy relation (2.1) is equivalent with:

$$\text{(H)} \quad A'-A \in 2C_k \quad \text{or} \quad A'-m(A) \in 2C_k$$

$$\text{(EH)} \quad A'-A \in 2C_k \quad \text{or} \quad A'-m(A) \in 2C_k \quad \text{or} \quad A'-m^2(A) \in 2C_k .$$

This, plus some computations, permits computation of the number of conjugacy classes when the group C_k is finitely generated:

Proposition 2.3 - Suppose that $\mathrm{char}(k) \neq 3$ and that C_k is a finitely generated group of rank r with 2^t elements of order 2 or 1 ($t = 0,1,2$ as well known). Then the number of conjugacy classes of subfields is:

(general case) 2^{r+t}

(H) (here $r = 2s$ and $t = 1,2$) $2^{2s+b-1} + 2^{s+1} - 2^s$

(EH)(here $r = 2s$ and $t = 0,2$) $1 + \frac{1}{3}(2^{2s+t} - 1)$.

Remark 2.4 - If $k = \mathbb{R}$, the cases (H) and (EH) cannot occur. We have $t = 1,2$ and t is the number of connected components of $C_{\mathbb{R}}$. Then $2C_{\mathbb{R}}$ is the unit component, and the number of conjugacy classes is t.

Proposition 2.5 - Let $K = k(t)$ and $K_1 = k(t_1)$ be two subfields of genus 0 and index 2 of $k(C)$, and let us write $k(C) = k(s,t) = k(s_1,t_1)$ with $s^2 = P(t)$ and $s_1^2 = P_1(t)$ (cf. I.2). Then K and K_1 are conjugate iff there exists an element θ of $PGL(2,k)$, $\theta(x) = \frac{ax+b}{cx+d}$, and $e \in k^*$ such that

(2.6) $$(ct+d)^4 P_1\left(\frac{at+b}{ct+d}\right) = e^2 P(t).$$

If $k(t)$ and $k(t_1)$ are conjugate, let σ be an automorphism of $k(C)$ such that $\sigma(k(t_1)) = k(t)$. Then $\sigma(t_1) = \frac{at+b}{ct+d}$ with $a,b,c,d \in k$, $ad-bc \neq 0$, and $\sigma(s_1)^2 = P_1(\frac{at+b}{ct+d})$. Since s and $\sigma(s_1)$ generate the same quadratic extension of $k(t)$, $\sigma(s_1)^2$ and s^2 differ by a square in $k(t)$, whence also $(ct+d)^4 P_1(\frac{at+b}{ct+d})$ and $P(t)$. But, since the two elements are squarefree quartic polynomials, their quotient must be a constant e^2.

Conversely, if (2.5) holds, let σ' be the k-isomorphism of $k(t_1)$ onto $k(t)$ defined by $\sigma'(t_1) = \frac{ab+b}{ct+d}$. Then the element $s' = \frac{es}{(ct+d)^2}$ satisfies the equation $s'^2 = P_1(\sigma'(t_1)) = \sigma'(P_1(t_1))$. Therefore σ' may be extended to an automorphism σ of $k(C)$ such that $\sigma(s_1) = s'$. Q.E.D.

Remark 2.7. - Let $P(X)$ and $P_1(X)$ be two squarefree quartics over k for which there exist $a,b,c,d,e \in k$, $ad-bc \neq 0$, $e \neq 0$, such that

(2.8) $$(cX+d)^4 P_1\left(\frac{aX+b}{cX+d}\right) = e^2 P(X) .$$

Then the quadratic extensions of $k(X)$ defined by $y^2 = P(X)$, $y_1^2 = P_1(X)$ are k-isomorphic. In other words, there exists an elliptic function field $k(C)$ and 4 elements s,b,s_1,t_1 of $k(C)$ such that $s^2 = P(t)$, $s_1^2 = P_1(t_1)$, $k(C) = k(s,t) = k(s_1,t_1)$.

§III - Classification of quartic polynomials

Proposition 2.5 and remark 2.7 lead us to study the action of PGL (2,k) on the squarefree quartics over k. Given such a polynomial $P(x)$ and an element θ of $PGL(2,k)$, $\theta(x) = \frac{ax+b}{cx+d}$, we denote by ${}^{\theta}P$ the polynomial defined by

(3.1) $$(cx+d)^4 P(\theta(x)) = P^*(a,c)\, {}^{\theta}P(x)$$

where P^* is the homogenized (artificial vitamins added!) polynomial of P (i.e. $P^*(x,y) = y^4 P(\frac{x}{y})$). The polynomial ${}^{\theta}P$ does not change if a,b,c,d are multiplied by the same constant, and is monic together with P

Remarks 3.9 - The polynomial $(cx+d)^4 P(\theta(x))$ depends, up to 4th powers, upon the choice of a,b,c,d. If $P^*(a,c) = 0$, it means that $P(x)$ has the rational root $\frac{a}{c}$; then $(cx+d)^4 P(\theta(x))$ is a cubic polynomial. Remark 2.2 shows that $P(x)$ and $P^*(a,c)\,{}^{\theta}P(x)$ give rise to conjugate subfields of the same elliptic function field.

We will study in this § the equivalence $P_1 = {}^{\theta}P$ of two monic quartics P, P_1. Later, it will be necessary to study also the effect of the leading coefficient $P^*(a,c)$, which, for our purposes acts through its class modulo squares. For the time being, only the first part of this program has been fairly well treated by us.

Proposition 3.3 - Let P and P_1 be two monic squarefree quartics over k, Σ and Σ_1 their sets of roots (in some algebraic closure of k). For P and P_1 to be equivalent under $PGL(2,k)$ (i.e. $P = {}^{\theta}P_1$), the conjunction of the following conditions is necessary and sufficient:

a) $k(\Sigma) = k(\Sigma_1)$

b) there exists a bijection $\theta' : \Sigma \to \Sigma_1$ which preserves the cross ratio and commutes with the Galois group Γ of $k(\Sigma)$ over k (i.e. $\theta'\varphi = \varphi\theta'$ for all $\varphi \in \Gamma$).

The necessity is clear. Conversely, a) and the equality of the cross ratios show that θ' may be extended to a fractional linear transformation $\theta : t \mapsto \frac{at+b}{ct+d}$ of $k(\Sigma)$, with coefficients a,b,c,d in $k(\Sigma)$. For any $\varphi \in \Gamma$, let θ^{φ} be the fractional linear transformation obtained by applying φ to the coefficients of θ. Then the condition $\theta'\varphi = \varphi\theta'$ means that we have $\theta^{\varphi}(t) = \theta(t)$ for any $t \in \Sigma$, whence $\theta^{\varphi} = \theta$ since Σ has 4 elements. Normalizing the coefficients a,b,c,d (by $c = 1$, or by $d = 1$ if $c = 0$), we conclude that they are in k, i.e. $\theta \in PGL(2,k)$.   Q.E.D.

Given a monic quartic polynomial $P(t)$ over k, with roots t_1, t_2, t_3, t_4, we are thus led to introduce its type, which comprises:

- the list of the degrees of the irreducible factors of P (e.g. (1,1,2), (2,2), (4))
- if necessary, a symbol giving information about the nature of the extension $k(t_1,t_2,t_3,t_4)$, in particular about its Galois group.

This decomposition field is denoted here by k'. In the following list of types we give the structure of the field $k(r)$ generated by any one of the six values of the cross ratio $r = (t_1, t_2, t_3, t_4)$, but, in a first approximation, the results about $k(r)$ exclude the cases (H) and (EH). In some cases, certain of the 6 values of r (called here the "preferred values", PCR) are easier to handle and permit to put the commutation condition of prop. 3.3 in a handy form. Of course, equivalent polynomials have the same type.

1) Type (1,1,1,1) - Here $k' = k(r) = k$. No PCR. The commutation condition is void.

2) Type (1,1,2) - Here k' is quadratic. We take $r = (t_1, t_2, t_3, \bar{t}_3)$ $(t_1, t_2 \in k)$ and its inverse as PCR. Two polynomials of type (1,1,2) are equivalent iff their PCR's are equal (or inverse); then there are 4 bijections θ' (of prop. 3.3).

3) Type (2,2)S (S for "same" quadratic extension) - Here k' is quadratic and $k(r) = k$. We take $r = (t_1, \bar{t}_1, t_2, \bar{t}_2)$ and its inverse as PCR. Two polynomials of type (2,2)S are equivalent iff their PCR's are equal or inverse; then there are 8 bijections θ'

4) Type (1,3)C (C for "cyclic")- Here k' is a cubic extension of k, with a 3-cyclic Galois group, and $k(r) = k'$. Having chosen a generator φ of the Galois group, we take $r = (t_1, t_2, t_2^{\varphi}, t_2^{\varphi^2})$ $(t_1 \in k)$ as PCR, and other ones being $1 - \frac{1}{r}$ and $\frac{1}{1-r}$ (exchange of φ and φ^2 inverts the PCR's). Two polynomials of type (1,3)C are equivalent iff their sets of PCR's are equal; then there are 3 bijections θ'.

5) Type (2,2)D (D for "distinct" quadratic extensions) - Here k', composite of two quadratic extensions, is a quartic extension, with a Klein group as Galois group; and $k(r)$ is the third quadratic subfield of k'. We take the PCR's in the form $r = (t_1, t_2, t_3, t_4)$ where (t_1, t_2) and (t_3, t_4) are pairs of conjugates over k; we get two of them, r and r', conjugate over k and such that $rr' = 1$. Two polynomials of type (2,2)D are equivalent iff their PCR's are equal or inverse; then there are 4 bijections θ'

6) Type (4)C (C for "cyclic") - Here k' is a cyclic extension of degree 4, and $k(r)$ is its unique quadratic subfield. Having chosen a generator φ of the

Galois group, we take $r = (t_1, t^{\varphi^2}, t_1^{\varphi}, t_1^{\varphi^3})$ as PCR (exchange of φ and φ^3 changes r into $\frac{1}{r}$); We get two of them, r and r', conjugate over k and such that $rr' = 1$. Two polynomials of type (4)C are equivalent iff their PCR's are equal or inverse; then there are 4 bijection θ'.

7) Type (4)K (K for "Klein") - Here k' is a quartic extension, with a Klein group $\{1, \varphi, \varphi', \varphi\varphi'\}$ as Galois group. Then $k(r) = k$. Having listed as above the elements of the Galois group, we take $r = (t_1, t_1^{\varphi}, t_1^{\varphi'}, t_1^{\varphi\varphi'})$ as PCR, and this value is unique. Two polynomials of type (4)K are equivalent iff they have the same PCR; then there are 4 bijections θ'.

8) Type $(1,3)\mathcal{S}_3$ - Here k' is an extension of degree 6 with $\mathcal{S}_3$ as Galois group, and $k(r) = k'$. No PCR. If two polynomials of type $(1,3)\mathcal{S}_3$ are equivalent, the bijection θ' of prop. 3.3 is unique.

9) Type (4)8 - Here k' is an extension of degree 8, and its Galois group is a 2-Sylow subgroups of $\mathcal{S}_4$. The field $k(r)$ is the quadratic extension of k corresponding to the Klein-subgroup of $Gal(k'/k)$ which is invariant in $\mathcal{S}_4$. No PCR. If two polynomials of type (4)8 are equivalent, there are 2 bijections θ' (the commutant of $Gal(k'/k)$ in $\mathcal{S}_4$ being its center, a group of order 2).

10) Type $(4)\mathfrak{A}_4$ - Here k' is an extension of degree 12, with the alternating group $\mathfrak{A}_4$ as Galois group, and $k(r)$ is the cubic cyclic extension of k corresponding to the invariant Klein subgroup of order 4 of $\mathfrak{A}_4$. No PCR. If two polynomials of type (4) $\mathfrak{A}_4$ are equivalent, the bijection θ' is unique (the commutant of $\mathfrak{A}_4$ in $\mathcal{S}_4$ being $\{1\}$).

11) Type $(4)\mathcal{S}_4$ - Here k' is an extension of degree 24, with $\mathcal{S}_4$ as Galois group, and $k(r)$ is the sextic extension of k corresponding to the invariant Klein subgroup of $\mathcal{S}_4$. No PCR. If two polynomials of type $(4)\mathcal{S}_4$ are equivalent, the bijection θ' is unique (since $\mathcal{S}_4$ has "no center").

We now return to elliptic fields. Given a subfield $K = k(t)$ of genus 0 and index 2 of an elliptic function field $k(C)$, the quartic polynomial $P(t)$ of (1.2) is determined up to equivalence, so that we can talk about the type of the subfield K. Conjugate subfields have the same type. Since the field $k(r)$ of the cross ratio is the same for all subfields K, the types which can occur in the same field $k(C)$ are not arbitrary. <u>Excluding the cases</u> (H) <u>and</u> (EH) (in which $k(r)$ gives little information), the above list shows that the following types may occur simultaneously in a given $k(C)$.

<u>Classification</u> (3.4) - a) (1,1,1,1), (2,2)S, (4)K

b) (1,1,2), (2,2)D, (4)C, (4)8

c) (1,3)C , (4)$\mathfrak{A}_4$

d) (1,3)$\mathcal{S}_3$, (4)$\mathcal{S}_4$

If k is a <u>finite field</u>, many types are excluded, and there remain only:

<u>Classification</u> (3.5) - a) (1,1,1,1), (2,2)S

b) (1,1,2), (4)C

c) (1,3)C

<u>Proposition</u> 3.6 - <u>If the types of two subfields</u> K, K_1 <u>of</u> $k(C)$ <u>contain both a</u> "one", <u>these subfields are conjugate</u> (with respect to a translation), <u>and the types are the same</u>.

In fact, consider the involutions $\tau: M \to -A-M$ and $\tau_1: M \to -A_1 - M (A, A_1 \in C_k)$ corresponding to K and K_1. The hypothesis means that there are points P, P_1 of C_k such that $2P = -A$, $2P_1 = -A_1$. If σ denotes the translation $M \mapsto (P_1 - P) + M$, one checks that $\tau_1 = \sigma \tau \sigma^{-1}$. Q.E.D.

<u>Corollary</u> 3.7 - <u>The number of "ones" in this type is the number of points of order</u> 1 <u>or</u> 2 <u>in</u> C_k. (Look at the cubic equation for C).

Thus, at least if k is <u>finite</u>, we can list the conjugacy classes in $k(C)$ and their types (cases (H) and (EH) temporarily excluded).

<u>Classification</u> 3.8:

a) 4 classes, one of type (1,1,1,1), 3 of type (2,2)S

b) 2 classes, one of type (1,1,2), one of type (4)C

c) Single class, of type (1,3)C

<u>Remark</u> 3.9 <u>about</u> $\mathbb{R}$ - Using remark 2.4, we have the following classification over $\mathbb{R}$:

a) Two classes, one of type (1,1,1,1), one of type (2,2)S

b) Single class of type (1,1,2) .

IV - Computational method when type (1,1,1,1) occurs.

Consider the elliptic curve with equation

$$y^2 = (x-a)(x-a')(x-a'') \qquad (a,a',a'' \in k_1 \text{ distinct})$$

The subfield $k(x)$ of $k(C)$ is of type (1,1,1,1). If (u,v) is a point of C_k, the corresponding subfield is $k(t)$ with $y-v = t(x-u)$. Set $b = u-a, b' = u-a'$, $b'' = u-a''$. A straight-forward computation shows that we can write $k(C) = k(s,t)$ with $s^2 = P(t)$ and

$$P(t) = t^4 - 2At^2 + 8vt + A^2 - 4B$$

where A,B,C are the elementary symmetric functions

$$b+b'+b'',\ bb'+bb''+b'b'',\ bb'b'' \quad \text{of } b,b',b'' .$$

If we try to decompose $P(t)$ into a product of quadratics

(*) $$P(t) = (t^2+\alpha t+\beta)(t^2-\alpha t+\beta')$$

we get the conditions

$$\beta+\beta'-\alpha^2 = 2A \qquad \alpha(\beta-\beta') = 8v, \qquad \beta\beta' = A^2-4B .$$

We first suppose $v \neq 0$. Taking $\gamma = \beta+\beta'$ as main unknown, we get the equation

$$(\gamma+2A)(\gamma^2-4A^2+16B) - 16v^2 = 0$$

(of course, the cubic resolvent of $P(t)$. Now $\delta = \frac{1}{4}(\gamma+2A)$ is a better unknown since it satisfies the equation

$$\delta^3 - A\delta^2 + B\delta + C = (\delta-b)(\delta-b')(\delta-b'') = 0 .$$

Thus the values of $\beta+\beta'$ are $2(b-b'-b'')$, $2(b'-b-b'')$ and $2(b''-b-b')$. Since we have the values of $\beta+\beta'$ and $\beta\beta'$, β and β' are the roots of a quadratic equation over k, the discriminant of which happens to be $4b'b'' = 4(u-a')(u-a'')$ (if we take $\beta+\beta' = 2(b-b'-b'')$). Since $(u-a)(u-a')(u-a'') = v^2$, this discriminant is a square in k iff $u-a$ is a square in k. Thus $P(t)$ decomposes over k into a product of two quadratics iff $u-a$, or $u-a'$, or $u-a''$ is a square in k.

At any rate, the product $(\alpha^2 - 4\beta)(\alpha^2 - 4\beta')$ of the discriminants of the factors of $P(t)$ is $16(a' - a'')^2$ (if we take $\beta + \beta' = 2(b - b' - b'')$), thus these factors have their roots in the same quadratic extension of $k(\sqrt{u-a})$. A further computation shows that this extension is $k(\sqrt{u-a}, \sqrt{u-a'})$.

Remark 4.1 - In general, the cubic resolvent of a quartic squarefree polynomial $B(t)$ over k has the following geometric interpretation: we choose a leading coefficient $e \in k$ such that the elliptic curve $s^2 = eB(t)$ has a k-rational point P. The involution with respect to $-2P$ gives a subfield of $k(t,s)$, the type of which contains a "one"; the corresponding quartic polynomial decomposes into a linear factor and a cubic one, and this cubic factor is essentially the resolvent of $B(t)$.

If $v = 0$ and, e.g., $u = a$, then $P(t) = t^4 + 2t^2(a' + a'' - 2a) + (a' - a'')^2$. A simple computation shows that the roots of $P(t)$ are $\pm\sqrt{u-a'} \pm \sqrt{u-a''}$. Furthermore, it decomposes over k into quadratics iff $a - a'$, or $a - a''$, or $(a-a')(a-a'')$ is a square in k.

Therefore we have the following result.

Proposition 4.2 - a) *If* $u-a$, $u-a'$ *and* $u-a''$ (resp. $a-a'$, $a-a''$ *for* $v=0$, $u=a$) *are squares in* k, *then* $P(t)$ *is of type* $(1,1,1,1)$ (*which means that* $(u,v) \in 2C_k$

b) *If only one of* $u-a$, $u-a'$, $u-a''$(resp. *of* $a-a'$, $a-a''$) *is a square in* k, *then* $P(t)$ *is of type* $(2,2,)S$.

c) *If none of* $u-a, u-a'$, $u-a''$ (resp. *of* $a-a'$, $a-a''$) *is a square in* k, *then* $P(t)$ *is of type* $(4)K$.

A straightforward computation of cross ratios (using (*) in case $v \neq 0$), plus a bit of Galois theory in case C), gives the additional information:

Proposition 4.3 - *If we chose* $\beta + \beta' = 2(b - b' - b'')$ (resp. $v = 0$, $u = a$), *a preferred value of the cross ratio for* $P(t)$ *is, in both cases* b) *and* c),

$$r = \frac{a'' - a}{a' - a} = (\infty, a, a', a'').$$

The other preferred value is $1/r$. Cases (H) and (EH) are not excluded here. Notice that all possible PCR's occur for suitable subfields of $k(C)$.

Complement 4.4 - The points (u,v) of C_k such that $u-a$ is a square form a subgroup D_a of C_k: in fact $k(x,y,\sqrt{x-a})$ is the function field of an elliptic curve E. The inclusion $k(C) \subset k(E)$ gives a morphism $f: E \to C$ and $f(E_k) = D_a$. With the origin "at infinity" on E, f is a group homomorphism. Proposition 4.2 shows that

$$(4.5) \qquad 2C_k = D_a \cap D_{a'} = D_{a'} \cap D_{a''} = D_{a''} \cap D_a .$$

Except possibly for $(a,0)$, $(a',0)$, $(a'',0)$, the points of $(D_a \cup D_{a'} \cup D_{a''}) - 2C_k$ correspond to the subfields of type (2,2)S. A computation of the index $(Da:2C_k)$ would be useful.

Example 4.6 - Finite fields - Let k be a finite field. Since $v^2 = (u-a)(u-a')(u-a'')$, one at least of $u-a$, $u-a'$, $u-a''$ (resp. of $a-a'$, $a-a''$, $(a-a')(a-a'')$) must be a square. Then proposition 4.2 shows that, even in cases (H) and (EH), the non-zero conjugacy classes are of type (2,2)S. Furthermore let $r = \frac{a''-a}{a'-a}$. Then prop. 4.3 gives

a) (General case) One class of type (1,1,1,1), 3 classes of type (2,2)S with PCR's respectively equal to $r, 1-r, \frac{r}{r-1}$.

b) (Case (H) with $i \in k$, 3 classes by prop. 2.3). One class of type (1,1,1,1), one class of type (2,2)S with PCR -1, one class of type (2,2,)S with PCR 2 .

c) (Case (EH) with $j \in k$, 2 classes by prop. 2.3). One class of type (1,1,1,1), one class of type (2,2)S with PCR $-j$.

Notice that, here, we have $(C_k:D_a) = (D_a:2C_k) = 2$.

Example 4.7 - Function fields as ground fields - Let $(C) y^2 = (x-a)(x-a')(x-a'')$ be defined over a field k_0. We take for k the function field $k = k_0(C')$ of another (or the same) elliptic curve C'. Let $\mathcal{E}$ be the group of k_0-homomorphisms of C' into C. As well known, C_k is the direct sum

$$(4.8) \qquad C_k = C_{k_0} \oplus \mathcal{E} .$$

The analogue is true for the subgroup D_a of complement (4.4):

(4.9) $$D_a = (D_a \cap C_{k_0}) \oplus (D_a \cap \mathcal{E})$$

(In fact, with the notation of 4.4, a point A of E_k corresponds to k_0-morphism $m_A: C' \to E$; we decompose m_A into a translation and a homomorphism, and we compose it with the homomorphism $f: E \to C$).

In particular, let us take $k_0 = \mathbb{F}_q$, $C' = C$ and assume that C has no complex multiplications (thus cases (H) and (EH) are excluded) so that $\mathcal{E}$, its ring of k_0-endomorphisms, is a $\mathbb{Z}$-module of rank 2. Then $C_k/2C_k$ is a group of order 16, whence 12 "transcendental" conjugacy classes besides the 4 "algebraic" ones. To find their types, (4.9) enables us to study only $D_a \cap \mathcal{E}$. One sees that neither the generic point (x,y), nor the Frobenius point (x^q,y^q) are in D_a (nor in $D_{a'}$, $D_{a''}$). But an amusing computation shows that the point $(x_1,y_1) = (x,y) + (x^q,y^q)$ is in these subgroups, whence in $2C_k$: the classical computation of x_1 as third root of a cubic equation gives

$$(x^q - x)^2(x_1 - a) = (y^q - y)^2 - (x^q - x)(x^q + x - a' - a'');$$

but, if we set $q = 2m+1$ and $z = y((x-a')^m(x-a'')^m - (x-a)^m)$, the computation of $(y^q - y)^2 - z^2$ shows that

$$(x^q - x)^2(x_1 - a) = z^2 \quad (!!) .$$

Thus, if we call f_q the Frobenius endomorphism of C, there is a $g \in \mathcal{E}$ such that $f_q = 2g - 1$.

> The fact that the characteristic roots of g are algebraic integers is equivalent with $4 | \mathrm{Card}\,(C_{k_0})$, nothing new since C is of type $(1,1,1,1)$ over $k_0 = \mathbb{F}_q$.

Examples lead us to think that the $\mathbb{Z}$-module $\mathcal{E}$ is generated by 1 and g. We have not yet computed the coordinates (x_2,y_2) of the point of C_k corresponding to g, nor checked that neither (x_2,y_2) nor $(x,y) + (x_2,y_2)$ belong D_a. We have the feeling that a computation showing, e.g., that $x_2 - a$ is a square in k would also show (by permutation) that $x_2 - a'$ and $x_2 - a''$ are also squares, i.e. that $(x_2,y_2) \in 2C_k$. This feeling leads us to <u>conjecture</u> that the 12 "transcendental" conjugacy classes are of type (4)K, and that each one of the 3 PCR's r, $1-r$, $\frac{r}{r-1}$ occurs 4 times in these classes.

§V - Isomorphism classes and conjugacy classes

Remark 2.2 gave an answer to what may be called the conjugacy problem: given two squarefree quartics P, P_1 over k, when do the equations $s^2 = P(t)$, $s_1^2 = P_1(t_1)$ define two conjugate subfields of the same elliptic function field $k(C)$? The isomorphism problem (over the algebraic closure $\bar{k}$ of k) is also easy: the curves defined by $s^2 = P(t)$ and $s_1^2 = P_1(t_1)$ are $\bar{k}$-isomorphic iff the roots of P and P_1 have the same set of cross ratios. More difficult seems to be the k-isomorphism problem, asking when these two curves are isomorphic over k.

If the types of P and P_1 both contain a "one", then prop. 3.6 shows that the k-isomorphism problem is the same as the conjugacy problem. Furthermore, a well known and quite explicit criterion can be given.

Proposition 5.1 - *Let k be a field of characteristic $\neq 3$*

a) *The elliptic curves* C: $y^2 = x^3 + bx + c$, and C': $y^2 = x^3 + b'x + c'$ *are k-isomorphic iff there exists* $d \in k^*$ *such that* $b' = d^4 b$ *and* $c' = d^6 c$ *(an isomorphism being* $(x,y) \mapsto (d^2 x, d^3 y)$

b) *The curves* $y^2 = e(x^3 + bx + c)$ *and* $y^2 = e'(x^3 + bx + c)$ *are k-isomorphic iff*

(general case, $c \neq 0$) e'/e *is a square*

(case (H), $c = 0$) $\pm e'/e$ *is a square*.

In fact, composing with a translation of C, it may be assumed that the k-isomorphism carries the point at infinity of C to the point at infinity of C'. Then it must have the form $(x,y) \mapsto (\alpha x, \beta y)$ and a simple computation gives the answers. Q.E.D.

Remark 5.2 - In characteristic 3, a non harmonic curve has a normal equation $y^2 = x^3 + ax^2 + c (a \neq 0)$. The curves $y^2 = x^3 + ax^2 + c$, $y^2 = x^3 + a'x^2 + c'$ are k-isomorphic iff there exists $d \in k^*$ such that $a' = d^2 a$ and $d = d^3 c$ (an isomorphism being $(x,y \mapsto (d^2 x, d^3 y)$. The analogue of b) holds. The harmonic case is not so handy.

Problem 5.3 - An interesting particular case of the k-isomorphism problem is the following one: P being a squarefree quartic over k, when are the curves

$$s^2 = eP(t) \qquad s^2 = e'P(t)$$

k-isomorphic? The condition "e'/e is a square" is sufficient in general, and also necessary if P is non-harmonic and if its type contains a "one" (prop.5.1,b)). In the non-harmonic case, it is also necessary if the intersection of the fields $k(t_i)$ (t_i: roots of P) is k itself: ground field extension to $k(t_i)$ introduces a "one" in the type of P and the square root of e'/e is in $k(t_i)$; this intersection property holds only in types (2,2)D, $(4)\mathcal{A}_4$, $(4)\mathcal{S}_4$.

Over a finite field k, the results of §§ III-IV show that the sufficient condition "e'/e a square" is also necessary for k-isomorphism. In fact, if P is of type (4)C, k-isomorphism means conjugacy (classification 3.8). If P is of type (2,2)S, then non-conjugate subfields have distinct PCR's (example 4.6).

But, in general, the condition "e'/e a square" is not necessary for k-isomorphism. Take a curve C of type (1,1,1,1) (neither (H), nor (EH)), $y^2 = (x-a)(x-a')(x-a'')$, defined over a finite field k_0, and the field $k = k_0(x,y,\sqrt{x-a})$ as new ground field. Then the "generic" point $G = (x,y)$ is in the subgroup denoted by D_a in complement 4.4, but neither in $D_{a'}$, nor in $D_{a''}$. Hence the corresponding subfield is of type (2,2)S. Also G is not in $C_{k_0} + 2C_k$ (in fact, $G = A + 2P$ gives, since $A \in C_{k_0}$ has finite order n, $nG = 2nP$, whence $G \in 2C_k$ since G has infinite order). Furthermore the 3 conjugacy classes of type (2,2)S given by points of C_{k_0} exhaust the 3 possible values of the PCR (example 4.6). Thus the monic quartics P, P_1 corresponding to G and to a suitable point of C_{k_0} have the same PCR and are equivalent under PGL(2,k)(3, 3). At the (cheap!) price of a fractional linear transformation, we may thus assume that $P = P_1$. Hence we can write $k(C) = k(s,t) = k(s't')$ with $s^2 = eP(t)$ and $s'^2 = e'P(t')$ $(e,e' \in k^*)$. But e'/e cannot be a square since, otherwise, the subfields would be conjugate.

The slight discrepancy in the harmonic case deserves a further study. It is due, roughly speaking, to the fact that a harmonic quartic admits more self-transformations under PGL(2,k) than a general quartic.

Counting classes over a finite field, 5.4 - Let k be the finite field, $\mathbb{F}_q$, q being prime to 6. It follows from prop. 5 a) that the number of k-isomorphism classes of elliptic curves $y^2 = x^3 + bx + c$ is

a) (general case, $b \neq 0$, $c \neq 0$) $2(q-2)$ classes

b) (case (H), $c = 0$) $\begin{cases} 4 \text{ classes if } q \equiv 1 \text{ mod. } 4 \\ 2 \text{ classes if } q \equiv 3 \text{ mod. } 4 \end{cases}$

c) (case (EH), $b = 0$) $\begin{cases} 6 \text{ classes if } q \equiv 1 \text{ mod. } 6 \\ 2 \text{ classes if } q \equiv 5 \text{ mod. } 6 \end{cases}$

Eg., we have 12 classes for $q = 5$, 18 classes for $q = 7$, 22 classes if $q = 11$.

Conjugacy classes of a given type are characterized by their PCR's and by the class modulo squares of the leading coefficient of the corresponding quartics. Thus, except possibly in cases (H) and (EH), the number of possible PCR's has to be multiplied by 2 to get the number of conjugacy classes.

Type (1,1,1,1)- Here the cross ratio r is in k, and must avoid the values 0,1, whence $q-2$ values. Prop. 5.1,b) gives:

(H) $\begin{cases} \text{one class if } q \equiv 3 \text{ mod. } 4 \ \ (-1 \text{ not a square}) \\ \text{two classes if } q \equiv 1 \text{ mod. } 4 \end{cases}$

(EH) (only if $q \equiv 1$ mod. 6) two classes

Other classes: $\frac{q-7}{3}$ if $q \equiv 1$ mod. 6, $\frac{q-5}{3}$ if $q \equiv 5$ mod. 6.

Type (1,1,2) - Here the PCR, r, is an element $\neq 1$ of $\mathbb{F}_{q^2}$ such that $r\bar{r} = 1$. There are q such elements, -1 being the only one to be its own inverse. Thus prop. 5.1., b) gives:

(H) $(r=-1)$ One class if $q \equiv 3$ mod. 4, 2 classes if $q \equiv 1$ mod. 4

(EH) (only if $q \equiv 5$ mod. 6) 2 classes.

Other classes: $q-1$ if $q \equiv 1$ mod. 6, $q-3$ if $q \equiv 5$ mod. 6.

Type (1,3)C - The easiest way is to proceed by differences.

Case (H) does not occur here.

(EH) (only if $q \equiv 1$ mod. 6) 4 classes

Other classes: $\frac{2(q-1)}{3}$ if $q \equiv 1$ mod. 6, $\frac{2(q+1)}{3}$ if $q \equiv 5$ mod. 6.

Type (2,2)S - Here the PCR is in k, whence $q-2$ values, -1 being its own inverse; the other ones give $\frac{q-3}{2}$ classes of monic polynomials. Then example 4 gives:

$$(H)\ \begin{cases} 3 \text{ classes (with } r=-1,\ r=2 \text{ and } r=\frac{1}{2}) & \text{if } q\equiv 3 \text{ mod. } 4 \\ 4 \text{ classes (two with } r=-1, \text{ two with } r=2) & \text{if } q\equiv 1 \text{ mod. } 4 \end{cases}$$

(EH) (only if $q\equiv 1$ mod. 6). Two classes (with $r=-j$).

Other classes: $q-7$ if $q\equiv 1$ mod. 6, $q-5$ if $q\equiv 5$ mod. 6.

Type (4)C - PCR as in type (1,1,2). Elimination of cases and prop.2.3 show that even in cases (H) and (EH), type (4)C can only occur together with type (1,1,2) in the same elliptic field. Thus:

(H) One class if $q\equiv 3$ mod. 4, two classes if $q\equiv 1$ mod. 4

(EH) (only if $q\equiv 5$ mod. 6) Two classes.

Other classes: $q-1$ if $q\equiv 1$ mod. 6, $q-3$ if $q\equiv 5$ mod. 6.

Remark 5.5 - Another way of computing the number of equivalence classes of monic polynomials is to make PGL $(2,\mathbb{F}_q)$ operate on a suitable set. This group has q^3-q elements.

For type (4)C, it operates on $\mathbb{F}_{q^4}-\mathbb{F}_{q^2}$. There are no fixed points, whence q orbits. If $t\in\mathbb{F}_{q^4}-\mathbb{F}_{q^2}$, then t^{q^2} is on the orbit of t, and (except if $(t,t^{q^2},t^q,t^{q^3})=-1$) t^q and t^{q^3} are in another orbit. Thus the total number of classes is $1+\frac{q-1}{2}=\frac{q+1}{2}$.

For type (2,2)S, we make $PGL(2,\mathbb{F}_q)$ operate on the set of pairs (u,v) of elements of $\mathbb{F}_{q^2}-\mathbb{F}_q$ such that $v\neq u,u^q$. This set has $(q^2-q)(q^2-q-2)$ elements, and, since there are no fixed points, we get $q-2$ orbits. Since $1, u+v, uv$ are linearly dependent over $\mathbb{F}_q$, there is an involution $\sigma\in PGL(2,\mathbb{F}_q)$ such that $\sigma(u)=v$, $\sigma(v)=u$, whence (u,v) and (v,u) are in the same orbit. But (u,v) and (u,v^q) are in distinct orbits, except if $(u,u^q,v,v^q)=-1$. Thus the number of classes is $1+\frac{q-3}{2}=\frac{q-1}{2}$.

For type (1,3)C, we make $PGL(2,\mathbb{F}_q)$ operate on $(\mathbb{F}_q\cup\infty)\times(\mathbb{F}_{q^3}-\mathbb{F}_q)$. There are no fixed points, whence $q+1$ orbits. The orbits of (a,t), (a,t^q) and (a,t^{q^2}) are distinct except in case (EH) (char. $\neq 3$).

§VI - Curves without rational points. The real case.

Here, our knowledge is more fragmentary. Let C be an elliptic curve defined over k without k-rational points. It is natural to introduce its Jacobian J, i.e. its group of divisor classes of degree 0. It is an elliptic curve, defined also over k, and isomorphic to C over the algebraic closure of k. Furthermore C is a principal homogeneous space over J: the elements of J operate as translations on C.

We recall that a field of genus 0 is either purely transcendental, or is the function field of a conic without rational points; it admits rational divisors of degree 1 or 2. Thus, for k(C) to possess subfields of genus 0 and index 2, it is sufficient that it admits a k-rational divisor of degree 2, and necessary that it admits a k-rational divisor of degree 4 (Remember for further use that the existence of a k-rational divisor D of degree $n > 0$ implies the existence of a positive k-rational divisor of degree n linearly equivalent to D).

If E is such a k-rational divisor of degree 4, the projective imbedding φ_E given by L(E) imbeds C in $\mathbb{P}_3$ and $\varphi_E(C)$ is a quartic curve, the intersection of two quadrics H and H′ defined over k. The linear pencil of quadrics $H + \lambda H' = 0$ contains 4 quadratic cones, obtained by writing that the discriminant of the quaternary form $H + \lambda H'$ vanishes (in thin air). We have

Proposition 6.1 - For k(C) to admit subfields of genus 0 and index 2 it is necessary and sufficient that C admits a k-rational divisor E of degree 4 such that one of the 4 quadratic cones containing $\varphi_E(C)$ be defined over k.

Necessity: a subfield of genus 0 and index 2 is the function field k(x,y) of a conic $F(x,y) = 0$. Then $k(C) = k(x,y,z)$ with $z^2 = G(x,y)$, G being a polynomial of degree ≤ 2. This gives an imbedding of C in 3-space, with affine equations $F(x,y) = z^2 - G(x,y) = 0$, and, if E denotes the divisor of poles of x (or y, or z), this imbedding is φ_E. The curve lies on the quadratic cone (or "cylinder") $F(x,y) = 0$.

Sufficiency: if S is the vertex of one of the quadratic cones defined over k the projection from S induces a k-morphism of degree 2 of $\varphi_E(C)$ onto a plane conic. Q.E.D.

Let us now view things in a more algebraic manner. For $k(c)$ to admit subfields of genus 0 and index 2, it is necessary and sufficient that $\mathrm{Aut}_k(C)$ contain, not only the group J_k of translations, but also a set of involutions. Here an involution on C is best defined by a linear system δ of divisors of degree 2 on C (or, what amounts to the same, by a divisor class of degree 2, called also δ in order to abuse language): the corresponding involution τ_δ is then defined by

(6.2) $$\tau_\delta(P) + P \in \delta \quad (\text{all } P \in C)$$

This involution τ_δ is defined over k iff the linear system δ is rational over k, i.e. if δ contains a divisor D such that $D^\varphi \sim D$ for every conjugate D^φ of D over k.

Then we have $D'^\varphi \sim D'$ for every $D' \in \delta$. This introduces cocycles which deserve a further study.

For any linear system δ rational over k, we denote by K_δ the field of invariants of τ_δ in $k(C)$.

Proposition 6.3 a) *Suppose* $\mathrm{Aut}_k(C)$ *contains only translations and involutions (e.g. if* J *has no complex multiplications). Then the subfields* K_δ *and* $K_{\delta'}$ *are conjugate iff* $\delta - \delta' \in 2J_k$ (here $\delta - \delta'$ is viewed as a divisor class of degree 0). *The set of conjugacy classes is equipotent with* $J_k/2J_k$.

b) K_δ *is purely transcendental over* k *iff* δ *contains a divisor of the form* $P + \bar{P}$, *where* P *is a point of* C *quadratic over* k *and* $\bar{P}$ *its conjugate*.

c) *Let* k' *be a quadratic extension of* k *such that* $C_{k'} \neq \emptyset$. *The set* $I_{k'}$ *of linear systems containing divisors of the form* $P + \bar{P}$ *with* $P \in C_{k'}$ *is a principal homogeneous space over a subgroup* $\mathcal{O}_{k'}$ *of* J_k *containing* $2J_k$ (i.e., *fixing* $\delta_0 \in I_{k'}$, *we have* $\delta \in I_{k'}$ *iff* $\delta - \delta_0 \in \mathcal{O}_{k'}$).

a) comes from an easy computation as in prop. 2. The fact that K_δ is purely transcendental iff it admits a valuation of degree 1, gives b). As to c), if the classes δ_i of $P_i + \bar{P}_i$ $(i = 0,1,2)$ belong to $I_{k'}$ $(P_i \in C_{k'})$, then the unique point $P_3 \in C$ such that $P_3 + P_0 \sim P_1 + P_2$ belongs to $C_{k'}$; calling δ_3 the class of $P_3 + \bar{P}_3$, we have $\delta_3 - \delta_0 = (\delta_1 - \delta_0) + (\delta_2 - \delta_0)$. Q.E.D.

Example 6.4. The real case.

An elliptic curve E defined over $\mathbb{C}$ can be put in the form $y^2 = x^3 + bx + c$. The number

$$(6.5) \qquad j'(E) = \frac{c^2}{4b^3 + 27c^2}$$

is an absolute invariant of E[1]. If b and c are real, then:

- $j'(E) > 0 \Rightarrow$ type $(1,1,2) \Leftrightarrow E_{\mathbb{R}}$ has one connected component $\Leftrightarrow$ the cross ratio is not real
- $j'(E) < 0 \Rightarrow$ type $(1,1,1,1) \Leftrightarrow E_{\mathbb{R}}$ has two components $\Leftrightarrow$ the cross ratio is real
- $j'(E) = 0$, harmonic case, one or two components.

Let now C be an elliptic curve defined over $\mathbb{R}$ without real points. It admits a $\mathbb{R}$-rational divisor D of degree 2, whence involutions defined over $\mathbb{R}$. Using $L(D)$ and $L(2D)$, we see that $\mathbb{R}(C) = \mathbb{R}(x,y)$ with $y^2 + P_4(x) = 0$, $P_4(x)$ being a monic quartic without real roots. The cross ratio $r = (\alpha,\bar{\alpha},\beta,\bar{\beta})$ of these roots is real. Since the Jacobian J of C has the same cross ratios as C, we must have

$$(6.6) \qquad j'(C) = j'(J) \leq 0$$

and J has type $(1,1,1,1)$, at least in the non-harmonic case. Then, by prop.6.3,a) there are two conjugacy classes of subfields of genus 0 and index 2.

A real fractional linear transformation brings the equation of C to the normal form

$$(6.7) \qquad y^2 + (x^2+1)(x^2+\lambda^2) = 0 \quad (\lambda \text{ real}, \neq 0,1,-1)$$

and we see two subfields of genus 0 and index 2:

- $\mathbb{R}(x)$, purely transcendental
- $\mathbb{R}(u,v)$ with $u = \frac{y}{x}$, $v = x + \frac{\lambda}{x}$, $u^2 + v^2 + (\lambda-1)^2 = 0$, not purely transcendental.

---

[1] It is a simple $PGL(2,\mathbb{Q})$ transform of the classical modular invariant.

Thus one of the conjugacy classes is formed by the purely transcendental subfields, the second by the subfields without valuations of degree 1. Since these two types of subfields exist in the harmonic case, J is also of type (1,1,1,1) in that case.

Complement 6.8 - Real curves with given absolute invariant j'.

As seen in 6.4, a real curve with invariant $j' > 0$ carries real points, thus can be put in the form $y^2 = x^3 + bx + c$. Since b^3/c^2 is fixed, prop. 5.1 shows that there are two $\mathbb{R}$-isomorphism classes of such curves, corresponding to opposite values of c.

For $j' < 0$ we still have two $\mathbb{R}$-isomorphism classes of curves carrying real points. In this case the 6 values of the cross ratio r are real, 2 are < 0, 2 are in $]0,1[$ and 2 are > 1; this set of 6 values is uniquely determined by j'. Since a cross-ratio for the curve with equation (6.7) is $\left(\frac{\lambda-1}{\lambda+1}\right)^2$, only the 4 positive values of r can be used to get a real λ, and each one gives two values of λ. Inverse values of r give opposite values for λ, whence the same λ^2. We thus get 4 possible values for λ^2 when j' is given. If one of them is denoted by μ^4, a straightforward computation shows that the other ones are

$$\frac{1}{\mu^4}, \left(\frac{1-\mu}{1+\mu}\right)^4, \left(\frac{1+\mu}{1-\mu}\right)^4 .$$

Since two curves corresponding to inverse values of λ^2 are $\mathbb{R}$-isomorphic, we have to see whether the curves C and C' with $\lambda^2 = \mu^4$ and $\lambda^2 = \left(\frac{1+\mu}{1-\mu}\right)^4$ are $\mathbb{R}$-isomorphic. If they were, their equations would, by 6.4, correspond to conjugate purely transcendental subfields of $\mathbb{R}(C)$, and a fractional linear transformation σ defined over $\mathbb{R}$ would carry the set $\{i,-i,i\mu^2,-i\mu^2\}$ of branch points onto $\{i, -i, i\left(\frac{1+\mu}{1-\mu}\right)^2, -i\left(\frac{1+\mu}{1-\mu}\right)^2\}$; since it carries then the imaginary axis onto itself, σ is necessarily of the form $\sigma(x) = dx$ or $\sigma(x) = \frac{e}{x}$ $(d, e \in \mathbb{R})$, and this happens only if $\mu = 1 \pm \sqrt{2}$, harmonic case. Thus we have two $\mathbb{R}$-isomorphism classes of curves with given $j' > 0$ and without real points.

For $j' = 0$, harmonic case, we still have two $\mathbb{R}$-isomorphism classes of curves with real points, $y^2 = x^3 \pm bx$ $(b > 0)$, one of type (1,1,1,1) and one of type (1,1,2). The curves without real points, in form (6.7), correspond to

$\left(\frac{\lambda-1}{\lambda+1}\right)^2 = 2$, i.e. to $\lambda = -3 \pm 2\sqrt{2}$, $\lambda^2 = 17 \pm 12\sqrt{2}$, two inverse values; thus they are $\mathbb{R}$-isomorphic. So we have proved:

Proposition 6.9 - Given a real value of the absolute invariant j', there are:

a) For $j' > 0$, two $\mathbb{R}$-isomorphism classes of real curves, both with real points.

b) For $j' < 0$, two $\mathbb{R}$-isomorphism classes of curves with real points, and two $\mathbb{R}$-isomorphism classes of curves without real points.

c) For $j' = 0$, two $\mathbb{R}$-isomorphism classes of curves with real points, and one $\mathbb{R}$-isomorphism class of curves without real points.

Now, for a given value of j', those curves without real points will, of course, have a Jacobian curve which is an elliptic curve with real points and the same value of j'. The situation is described by the cohomology group $H^1(\mathfrak{G}; J_{\mathbb{C}})$ where $\mathfrak{G}$ is the Galois group of $\mathbb{C}/\mathbb{R}$ and $J_{\mathbb{C}}$ is the group of complex points on the curve. For $j' > 0$, Proposition 6.9 easily shows that $H^1(\mathfrak{G}; J_{\mathbb{C}}) = 0$ and for $j' = 0$, $H^1(\mathfrak{G}; J_{\mathbb{C}}) = 0$ or is cyclic of order 2 depending upon whether the curve with real points is of type (1,1,2) or (1,1,1,1), respectively. When $j' < 0$, it is conceivable that, for the two isomorphism classes of real Jacobians, $H^1(\mathfrak{G}; J_{\mathbb{C}})$ might in one case be 0 and in the other cyclic of order 3. However, this cannot happen as the following argument shows. We consider the exact sequence, $0 \to J_{\mathbb{R}} \to J_{\mathbb{C}} \to E \to 0$ and the associated exact sequence for cohomology:

$$0 \to J_{\mathbb{R}}^{\mathfrak{G}} \to J_{\mathbb{C}}^{\mathfrak{G}} \to E^{\mathfrak{G}} \to H^1(\mathfrak{G}; J_{\mathbb{R}}) \to H^1(\mathfrak{G}; J_{\mathbb{C}}) .$$

Since $\mathfrak{G}$ acts trivially in $J_{\mathbb{R}}$ and $J_{\mathbb{C}}^{\mathfrak{G}} = J_{\mathbb{R}}$ we deduce the exact sequences:

$$0 \to E^{\mathfrak{G}} \to H^1(\mathfrak{G}; J_{\mathbb{R}}) \to H^1(\mathfrak{G}; J_{\mathbb{C}}) .$$

Now, the group of 1-cocycles of $\mathfrak{G}$ on $J_{\mathbb{R}}$ is simply $\mathrm{Hom}_{\mathbb{Z}}(\mathfrak{G}; J_{\mathbb{R}})$ and the group of 1-coboundaries is trivial since $\mathfrak{G}$ acts trivially. Finally an argument exactly imitating Kummer theory shows that $E = J_{\mathbb{R}}/2J_{\mathbb{R}}$. Instead of using the assumption that the ground field contains all n^{th} roots of unity one uses the fact that $J_{\mathbb{R}}$ contains all the "halves of 0", that is to say, the four points P_i such that

$2P_i = 0$ (remember we are in type (1,1,1,1).) Finally we note that $J_{\mathbb{R}}/2J_{\mathbb{R}}$ is of order 2 while $\mathrm{Hom}_{\mathbb{Z}}(\mathfrak{G}, J_{\mathbb{R}})$ is of order 4 (Klein 4) and $H^1(\mathfrak{G}; J_{\mathbb{C}})$ is of at most order 3 by part (b) of Proposition 6.9. Hence since 2 does not divide 3 we see that $H^1(\mathfrak{G}; J_{\mathbb{C}})$ is of order 2 for each real Jacobian. We have proved:

Proposition 6.10 - *Let* C *be an elliptic curve defined over* $\mathbb{R}$ *and having the absolute invariant* j'. *If* $J_{\mathbb{C}}$ *is the group of complex points in* C *then:*

a) *if* $j' > 0$, *then* $H^1(\mathfrak{G}; J_{\mathbb{C}}) = 0$

b) *if* $j' < 0$, *then* $H^1(\mathfrak{G}; J_{\mathbb{C}})$ *is of order* 2.

c) *if* $j' = 0$, *then* $H^1(\mathfrak{G}; J_{\mathbb{C}})$ *is zero or of order* 2 *depending on whether* C *is of type* (1,1,2) *or* (1,1,1,1), *respectively*.

For the time being, we know very little beyond the real case. A criterion for the existence of subfields of genus 0 and index 2, more algebraic than (6. 1) would be useful. The following problems, in which the existence of such subfields of k(C) is assumed, seem open:

a) Can they all be purely transcendental over k without C admitting k-rational points?

b) Can none of them be purely transcendental? (this means that C has no quadratic points over k).

Non conjugate purely transcendental subfields occur: take $k = \mathbb{Q}$, and C as the plane quartic curve $y^2 + (x^2+x)(x^2+3)$. Then k(x) and k(t), with $y = t(x^2+2)$, have index 2, but are not conjugate since the splitting fields of the corresponding quartic polynomials are distinct.

c) Given an equation for an elliptic curve with k-rational points, is there an explicit way of deriving equations for elliptic curves without k-rational points for which the original curve is the Jacobian? It seems to be possible to do this if we are content to consider curves with quadratic points.

§VII - Other open questions.

Even in the case of curves with rational points, various problems seem open:

a) Only part of our results is valid in cases (H) and (EH). A systematic treatment of these cases would be useful.

b) From time to time, characteristic 3 has been excluded (especially the harmonic case of characteristic 3, in which $Aut_k(C)$ can be quite large). It would be good to fill the gap.

c) The case of characteristic 2 probably requires quite different methods. E.G., instead of 4 branch points of $k(C)$ over K, we have only 1 or 2 of them (with wild ramification), so that the reasonings based upon cross-ratios fail.

d) The splitting fields (over k) of the various quartic polynomials, associated with a given elliptic curve C, deserve a closer study. In particular, if k is a local or a global field, their ramification over k could be studied.

THE GROUP OF DIVISIBILITY AND ITS APPLICATIONS

Joe L. Mott*, Department of Mathematics
Florida State University, Tallahassee, Florida 32306

INTRODUCTION

Associated with any integral domain D is a partially ordered directed group G(D). This group, the set of non-zero principal fractional ideals of D with $aD \leq bD$ if and only if aD contains bD, is called the group of divisibility of D. If K^* denotes the multiplicative group of the quotient field of D and U(D) the group of units of D, then G(D) is order isomorphic to $K^*/U(D)$, where $aU(D) \leq bU(D)$ if and only if $b/a \in D$.

Certainly, Dedekind [9] recognized that the study of divisibility by elements of D amounts, essentially, to the study of G(D). Indeed, G(D) is lattice ordered if and only if D is a GCD domain [16, p. 267]. The missing ingredient in Kummer's attempted proof of Fermat's Last Theorem was that for some algebraic number fields G(D) need not be lattice ordered. To overcome this obstacle, Dedekind created the theory of ideals. Essentially, his idea was to embed G(D) into some lattice ordered group. In the cases considered by Dedekind, the set F of all non-zero fractional ideals of D formed an ℓ-group.

In this context one can trace several other developments in ring theory. Since F need not be a group, Prüfer [29], Krull [20], Lorenzen [21], van der Waerden [34, §105] and others studied other systems of ideals contained in F and containing G(D). Krull [20, p. 665] observed that an integral domain D is completely integrally closed if and only if the set of v-ideals $V(D)$ form a group, and, in this case, $V(D)$ is a complete ℓ-group. Prüfer considered domains where the set of finitely generated fractional ideals form a group (indeed, an ℓ-group). Lorenzen discovered necessary and sufficient conditions for an abelian group to be embedded in an ℓ-group

*This work was supported by National Science Foundation Grant GP-19406.

(Dieudonné [10] later gave an elegant proof without using systems of ideals). Krull [19, p. 165] proved that for a valuation ring D, G(D) is merely the value group. In this case, ideal theoretic properties of D are easily derived from corresponding properties of G(D), and conversely.

Subsequently, several mathematicians have derived, and have effectively employed, a method involving the group of divisibility to solve certain ring theoretic problems. This method involves three general steps. First, formulate the ring theoretic problem in terms of the group of divisibility; second, solve the problem there; then pull back the solution, whenever possible, to the integral domain. Apparently, Lorenzen [21] originally used this approach to solve a problem of Krull, and Nakayama [25] applied it to disprove conjectures of Krull [20] and Clifford [5]. Two papers by Ohm [27, 28] greatly popularized the method, and recently Heinzer [14, 15] and Sheldon [32] employed it to negatively answer ring theoretic questions.

The main tool of this method is a sequence of theorems proved by Krull, Kaplansky, Jaffard, and Ohm. These theorems, when summarized, state that for any lattice-ordered group G, there is a Bezout domain D with G as its group of divisibility. Throughout the paper, this result will be designated as the Krull-Kaplansky-Jaffard-Ohm Theorem.

Authors have generally applied the method to construct counterexamples to conjectures in ring theory. They have stressed that the advantage of this process is the abundance of examples of O-groups whereas examples of integral domains are not as abundant. While not denying this aspect of the theory, I am interested in the opposite end of the scale. My primary purpose here is to show how several classical results in ordered group theory follow naturally from a corresponding ring theory result via the group of divisibility. The principal tool in this approach is the Krull-Kaplansky-Jaffard-Ohm Theorem. This theorem and other results in the paper give, among others, Lorenzen's embedding [21] of an abelian ℓ-group in a cardinal product of totally ordered groups, an embedding of an archimedean ℓ-group in a complete ℓ-group [7], and Ward's [33] and Birkhoff's [2] characterization of a cardinal sum of infinite cyclic groups.

1. DEFINITIONS AND NOTATIONS

The notation and terminology will essentially be the same as that of Ohm's paper [28]. In this paper, all groups are abelian; an 0-group is partially ordered, and an ℓ-group is lattice-ordered. A cartesian product of 0-groups G_λ is called the cardinal product (sometimes called the ordered direct product or a vector group) if $x = (x_\lambda) \geq y = (y_\lambda)$ if and only if $x_\lambda \geq y_\lambda$ for each λ. The cardinal product of the group G_λ will be denoted by $\Pi_\lambda G_\lambda$. An 0-group G is a subcardinal product (usually called a subdirect sum) of the groups G_λ if there is an 0-embedding ϕ of G into $\Pi_\lambda G_\lambda$ such that $p_\lambda\phi(G) = G_\lambda$ for each λ, where p_λ is the canonical projection map of $\Pi_\lambda G_\lambda$ onto G_λ. The cardinal sum of the groups G_λ, denoted by $\sum_\lambda G_\lambda$, is the subset of $\Pi_\lambda G_\lambda$ of all elements with finite support.

Let Z denote the group of integers under the natural order and let R denote the additive group of real numbers.

If $a_0, a_1, \ldots, a_n$ are elements of an 0-group G, $a_0 \in \sup(\inf_G \{a_1, \ldots, a_n\})$ means a_0 is an upper bound of the set of all lower bounds of $a_1, \ldots, a_n$ (in Ohm's notation $a_0 \geq \inf_G \{a_1, \ldots, a_n\}$). If G is an ℓ-group, let cup ($\vee$) and cap ($\wedge$) denote sups and infs. If $a, b \in G$ then $a||b$ means $a \not\geq b$ and $b \not\geq a$.

A semi-valuation of a field K is a map v of K^* onto an (additive) abelian 0-group G such that for all $x, y \in K^*$

(1) $v(xy) = v(x) + v(y)$

(2) $v(x + y) \in \sup(\inf_G \{v(x), v(y)\})$

(3) $v(-1) = 0$.

The group G is called the semi-value group of v.

The canonical example of a semi-valuation is the natural homomorphism $v: K^* \to K^*/U(D)$, where $U(D)$ is the group of units in a subring D of K, and the positive cone of $K^*/U(D)$ is the set of elements $dU(D)$, where $d \in D \setminus \{0\}$. In general, $K^*/U(D)$ is not a directed (filtered) group; in fact, $K^*/U(D)$ is directed if and only if K is the quotient field of D, and then $K^*/U(D)$ is the group of divisibility of D. (I usually prefer to think of the canonical semi-valuation of K^* onto the group of divisibility of D as the map v that maps $x \in K^*$ to the principal fractional ideal xD.)

If, conversely, v is a semi-valuation on K with semi-value group G, the set $D = \{x \in K^* \mid v(x) \geq 0\} \cup \{0\}$ is a subring of K and $K^*/U(D)$ is 0-isomorphic to G.

Recall, finally, that a Bezout domain is an integral domain for which finitely generated ideals are principal. The group of divisibility of a Bezout domain is lattice-ordered.

2. PRELIMINARY RESULTS

THEOREM 2.1. (Krull-Kaplansky-Jaffard-Ohm) *If G is a lattice-ordered abelian group, there is a Bezout domain D with group of divisibility order isomorphic to G.*

The evolution of this basic result required a period of more than thirty years. Krull [19, p. 164] made the first significant contribution, proving that for a totally ordered group G, there is a valuation ring D with G as its value group. The essentials of his proof can be described as a series of definite steps, roughly as follows.

Consider the group algebra $k[G]$ for any field k. Define w on $k[G]$ by $w(\sum \alpha_g g) = \inf \{g \mid \alpha_g \neq 0\}$. Extend w to the quotient field K of $k[G]$ by $w(f/h) = w(f) - w(h)$, where $f, h \in k[G]$. Observe that w is a semi-valuation (in this case, a valuation) on K. Next, let $D = \{x \in K^* \mid w(x) \geq 0\} \cup \{0\}$. Then, D is an integral domain with group of divisibility 0-isomorphic to G.

Jaffard, following the major steps of Krull's proof, first published the next generalization [16, p. 263] (although Kaplansky had made the same observation some twelve years earlier in an unpublished portion of his dissertation at Harvard University). For an ℓ-group G, Jaffard constructed a domain D with G as its group of divisibility. Then Ohm [27, p. 329] astutely observed that the domain D in this construction is necessarily a Bezout domain.

Realizations of an 0-group by totally ordered groups have received considerable attention. The question is this: When can a given 0-group be embedded as a subgroup of a cardinal product of totally ordered groups? The following theorem of Fuchs' [12] will be useful in obtaining realizations of groups of divisibility.

THEOREM 2.2. (Fuchs) *If an integral domain* D *is an intersection of valuation overrings* V_α *of* D, *then the group of divisibility* $G(D)$ *of* D *is a subcardinal product of the totally ordered groups* $G(V_\alpha)$.

The proof is straightforward; map the principal fractional ideal xD to $(xV_\alpha)_\alpha$ in the cardinal product of the groups $G(V_\alpha)$. On other occasions I shall refer to this embedding as *Fuchs' embedding*.

Theorem 2.2 has a partial converse (Proposition 2.3). The ideas, definitions, and results involved in the proof of Proposition 2.3 are all contained in Ohm's paper [28]. I include these results because of their usefulness throughout the paper.

DEFINITION. If B and C are 0-groups and σ is a homomorphism of B into C, then σ is a *V-homomorphism* if for all $b_0, b_1, \ldots, b_n \in B$ with $b_0 \in \sup(\inf_B \{b_1, \ldots, b_n\})$, then $\sigma(b_0) \in \sup(\inf_C \{\sigma(b_1), \ldots, \sigma(b_n)\})$. A V-embedding is a one-to-one V-homomorphism.

Note that a V-homomorphism is necessarily an 0-homomorphism.

In the sequel to Theorem 2.2, I shall assume that the group of divisibility $G(D)$ of an integral domain D is a subcardinal product of totally ordered groups G_λ in such a way that the map $v_\lambda = p_\lambda \phi v$ is a semi-valuation for each λ (here v is the canonical semi-valuation, ϕ is Fuchs' embedding, and p_λ is the projection of $\Pi_\lambda G_\lambda$ onto G_λ). Then $V_\lambda = \{x \in K^* \mid v_\lambda(x) \geq 0\} \cup \{0\}$ is a valuation overring of D and $D = \cap_\lambda V_\lambda$.

The following results concerning V-homomorphisms imply that each v_λ is a semi-valuation whenever ϕ is a V-homomorphism.

1) *Let* v *be a semi-valuation on* K *with value group* (B, P). *If* β *is a V-homomorphism of* B *into a partially ordered group* (C, P') *then* βv *is also a semi-valuation on* K *with value group* $(\beta(B), P' \cap \beta(B))$.

2) *The composition of two V-homomorphisms is a V-homomorphism.*

3) *If* B *and* C *are lattice-ordered groups and* β *is a 0-homomorphism of* B *into* C, *then* β *is a V-homomorphism if and only if* β *is a* ℓ*-map (that is,* β *preserves* infs *and* sups) *onto a sublattice of* C.

4) The projection map p_λ of the cardinal product of directed O-groups G_α onto G_λ is a V-homomorphism.

5) Suppose σ_λ is an O-homomorphism of C into an O-group G_λ. Let σ be the cartesian product of all σ_λ. Then σ is a V-homomorphism of C into the cardinal product $\Pi_\lambda G_\lambda$ if and only if σ_λ is a V-homomorphism for each λ.

Proposition 2.3 then follows from the above considerations.

PROPOSITION 2.3. If the group of divisibility $G(D)$ of a domain D is V-embedded as a subcardinal product of totally ordered groups G_λ, then D is an intersection of valuation rings V_λ, where $G(V_\lambda) = G_\lambda$.

An example of Jaffard

The group of divisibility of a domain must necessarily be a directed group. Jaffard [18] first published an example of a directed group that is not the group of divisibility of an integral domain.

Jaffard's example is the following: Let J be the subgroup of the cardinal sum of two copies of Z where $(a, b) \in J$ if and only if $a + b$ is even. Furthermore, the order on J is the induced partial order. The identity map is a V-embedding of J as a subcardinal product of $Z \oplus Z$. If, therefore, J is the group of divisibility of some domain D, then, by Proposition 2.3, D must be an intersection of two rank one discrete valuation rings. But, then, D is a Bezout domain (in fact, a PID) [24, p. 38], and J is lattice-ordered. The conclusion follows at once: J is not the group of divisibility of an integral domain.

Paul Hill showed me the following elementary proof. Consider the two elements (2, 2) and (3, 1) in J: If J is the group of divisibility of some domain D, then for certain elements a and b of D, $(2, 2) = v(a)$ and $(3, 1) = v(b)$, where v is the canonical semi-valuation. Then $v(a + b) = (c, d) \in \sup \{\inf_G \{v(a), v(b)\}\}$ is greater than (1, 1) and (2, 0) so that $c \geq 2$ and $d \geq 1$. If $c = 2$ then $d \geq 2$, and if $d = 1$ then $c \geq 3$. In either case, $v(a + b) \geq v(a)$ or $v(a + b) \geq v(b)$. Therefore, $(a + b)D$ is contained in aD or bD, and, hence, $bD \subseteq aD$ or $aD \subseteq bD$. But then, one of (2, 2) and (3, 1) must be greater than the other—a contradiction.

In this context, it is appropriate to mention a result in a paper by Cohen and Kaplansky [6]. I translate their result into the language of the group of divisibility. Recall that an atom is a minimal positive element of an 0-group.

THEOREM 2.4. (Cohen and Kaplansky) Suppose that G is a partially ordered group with positive cone P. Suppose, further, that (G, P) is not an ℓ-group and that P satisfies the descending chain condition. If (G, P) is the group of divisibility of some domain D, then P contains more than two atoms. Further, if P contains exactly three atoms a_1, a_2, and a_3, then $a_i + a_j = 2a_k$ for every possible permutation (i, j, k) of 1, 2, and 3.

Thus, Theorem 2.4 gives a simple method for constructing 0-groups (G, P) that cannot be groups of divisibility. For instance, it is easy to see that the set $P = \{n \in Z \mid n \geq 2 \text{ or } n = 0\}$ determines a directed partial order on Z and that P contains exactly two atoms. Jaffard's example J satisfies the d.c.c. on positive elements since J can be 0-embedded in the cardinal sum $Z \oplus Z$. Furthermore, J has exactly three atoms $a_1 = (1, 1)$, $a_2 = (2, 0)$ and $a_3 = (0, 2)$. Since $a_1 + a_2 \neq 2a_3$, J cannot be a group of divisibility.

3. AN ANALYSIS OF FUCHS' EMBEDDING

If an integral domain D is an intersection of valuation rings V_α, one might expect that Fuchs' embedding ϕ of $G(D)$ into the cardinal product of groups $G(V_\alpha)$ is a V-homomorphism. We shall presently see, however, that this is not the case.

First, recall some facts about $*$-operations [13, 26]. If F is a fractional ideal of D, the v-ideal of F, denoted by F_v, is the intersection of all principal fractional ideals containing F. The v-operation is the map $F \to F_v$. If D is an intersection of valuation rings V_α, the w-operation determined by the valuation rings $\{V_\alpha\}$ is the map $F \to F_w$, where $F_w = \cap_\alpha FV_\alpha$ is the w-ideal of F.

Next, observe that $b_0D \in \sup(\inf_{G(D)} \{b_1D, \ldots, b_nD\})$ means b_0D is contained in every principal fractional ideal that contains $b_1D, \ldots, b_nD$, that is, $b_0 \in (b_1, \ldots, b_n)_v$. In particular, if D is a Bezout domain, this means $b_0 \in (b_1, \ldots, b_n)D$, since $(b_1, \ldots, b_n)D$ is a principal ideal.

Suppose $D = \cap_\alpha V_\alpha$, where V_α is a valuation ring for each α. Suppose, further, that Fuchs' embedding is a V-homomorphism. Then, $b_0D \in \sup(\inf_{G(D)} \{b_1D, \ldots, b_nD\}$ implies $b_0V_\alpha \in \sup(\inf_{G(V_\alpha)} \{b_1V_\alpha, \ldots, b_nV_\alpha\})$ for each α. In other words: $b_0 \in (b_1, \ldots, b_n)_v$ implies $b_0 \in (b_1, \ldots, b_n)V_\alpha$ for each α. But then, $b_0 \in \cap_\alpha (b_1, \ldots, b_n)V_\alpha = (b_1, \ldots, b_n)_w$, and $(b_1, \ldots, b_n)_v \subseteq (b_1, \ldots, b_n)_w$. Therefore, $(b_1, \ldots, b_n)_v = (b_1, \ldots, b_n)_w$ since F_v contains F^* for each ideal F and each $*$-operation [13, p. 389]; thus, the v-operation is equivalent to the w-operation determined by the valuation rings V_α.

After verifying the reverse implications, we conclude the following theorem.

THEOREM 3.1. *Suppose* D *is an intersection of valuation rings* V_α. *Let* w *be the associated* $*$-*operation*, $w: F \to F_w = \cap_\alpha FV_\alpha$. *Then, Fuchs' embedding of* $G(D)$ *into the cardinal product of the groups* $G(V_\alpha)$ *is a V-homomorphism if and only if the v-operation is equivalent to* w.

COROLLARY 3.2. *If* D *is an intersection of valuation rings* V_α, *where each* V_α *is a quotient ring of* D, *then Fuchs' embedding is a V-homomorphism.*

Proof. See [13, p. 549].

Since the w-operation determined by valuation rings is "endlich arithmetische brauchbar" [13, p. 362], the embedding of Fuchs' is a V-homomorphism only if D is a v-domain [13, p. 391]. In fact, we have the following corollary.

COROLLARY 3.3. *The group of divisibility of* D *is V-embedded as a subcardinal product of totally ordered groups if and only if* D *is a v-domain.*

Proof. If D is a v-domain, then the v-operation is equivalent to some w-operation determined by a collection of valuation rings $\{V_\alpha\}$, where $D = \cap_\alpha V_\alpha$. Then, Theorem 3.1 implies $G(D)$ is V-embedded in $\Pi_\alpha G(V_\alpha)$.

Remarks

If D is a Krull domain, the only w-operation equivalent to the v-operation is the w-operation determined by D_{P_α} for all minimal primes P_α of D. For if V

is a valuation overring of D centered at some non-minimal prime ideal P of D, then $P_v = D$, $PV \neq V$, and P_v is the v-ideal of some finitely generated $A \subseteq P$ (see the proof of 36.3 in [13, p. 534]).

In particular, if $D = k[x, y]$, where k is a field, then $D = \cap_\alpha V_\alpha \cap V$, where V_α is D_{P_α} for a minimal prime P of D and V is some valuation ring centered on (x, y). Then $1 \in (x, y)_v$ but $1 \notin (x, y)V$ so that Fuchs' embedding of $G(D)$ into the cardinal product of all $G(V_\alpha)$ and $G(V)$ is not a V-homomorphism.

4. SOME CLASSICAL RESULTS

In 1939, Lorenzen [21] proved that an ℓ-group can be embedded in a cardinal product of totally ordered groups as a sublattice and as a subcardinal product. Since then, Birkhoff [2], Clifford [5], Jaffard [16, 17], Ribenboim [30, 31], and others reproved and refined this result. I prove it using the Krull-Kaplansky-Jaffard-Ohm Theorem. Subsequently, I shall prove other familiar results using the same approach.

THEOREM 4.1. (Lorenzen) *If G is a lattice-ordered group, then there is a family of totally ordered groups $\{G_\lambda\}$ and an embedding ϕ of G onto a sublattice of $\Pi_\lambda G_\lambda$. Moreover, $\phi(G)$ is a subcardinal product of the groups G_λ.*

Proof. Since G is lattice-ordered, there is a Bezout domain D with G as its group of divisibility. Thus, D is an intersection of valuation rings V_α, where each V_α is a quotient ring of D. By Corollary 3.2, G can be V-embedded as a subcardinal product of the totally ordered groups $G(V_\alpha)$. Since G is lattice-ordered, this V-embedding is a lattice homomorphism onto a sublattice by result 3 of §2.

COROLLARY 4.2. *If D is an integral domain with lattice-ordered group of divisibility, then D is integrally closed.*

Proof. By Theorem 4.1, $G(D)$ is V-embedded as a sublattice of $\Pi_\lambda G_\lambda$, where each G_λ is totally ordered. Then by Proposition 2.3, D is an intersection of valuation rings V_λ where $G(V_\lambda) = G_\lambda$. In particular, D is integrally closed.

A second consequence of the Theorem 2.1 is the following expanded version of a theorem of Ward [33] and Birkhoff [2] (I have added part 3 for completeness).

THEOREM 4.3. (Ward and Birkhoff) Suppose G is a lattice-ordered abelian group. These are equivalent.

1) G is a cardinal sum of copies of Z.

2) G satisfies the descending chain condition on positive elements.

3) G is the group of divisibility of a unique factorization domain.

Proof. Clearly 1) implies 2) and 3) implies 1). We show 2) implies 3). There is a domain D with G as its group of divisibility. Since G satisfies d.c.c. on positive elements, D satisfies a.c.c. on principal ideals. Hence, each non-zero non-unit of D is a finite product of irreducible elements. That D is a UFD will follow if we observe that an irreducible element p of D is prime.

Let v be the canonical semi-valuation from the quotient field of D onto G. Then, p irreducible in D implies $v(p)$ is a minimal positive element of G. Thus, to observe that p is prime, we need only see that if $v(p) \leq v(x) + v(y)$ where $v(x) \geq 0$ and $v(y) \geq 0$, then $v(p) \leq v(x)$ or $v(p) \leq v(y)$. But that is a well-known property of minimal positive elements in a lattice-ordered group [11, p. 71].

The next theorem apparently first appeared as an exercise in [3, p. 90, Ex. 31] (see also [13, p. 548]). The result will be an easy corollary of Theorem 4.3 using some well-known facts about Kronecker functions. (See [1], [20] or [13, 26].)

COROLLARY 4.4. (Bourbaki) If D is a Krull domain, the Kronecker function ring D^v is a principal ideal domain.

Proof. Since D is a Krull domain, $D = \cap_\alpha V_\alpha$, where each $V_\alpha = D_{P_\alpha}$ is a discrete rank one valuation ring and each non-zero non-unit of D is a non-unit in only finitely many of the valuation rings V_α (that is, $\{V_\alpha\}$ has finite character on D). Now the Kronecker function ring D^v is equal to $\cap_\alpha V_\alpha(X)$, each $V_\alpha(X)$ is rank one discrete and $\{V_\alpha(X)\}$ has finite character on D^v [7]. Therefore, $G(D^v)$ is embeddable in the cardinal sum of $G(V_\alpha(X))$ and $G(D^v)$ is lattice-ordered since

D^v is a Bezout domain. Hence, $G(D^v)$ satisfies the d.c.c. on positive elements. By Theorem 4.3, $G(D^v)$ is a cardinal sum of copies of Z and D^v is a UFD. Since D^v is Bezout, D is, in fact, a PID [13, p. 525].

Another accomplishment of Theorem 2.1 is the embedding of an archimedean lattice-ordered abelian group in a complete ℓ-group. MacNeille [22] showed that the classical method of forming Dedekind cuts can be applied to obtain an embedding of an arbitrary partially ordered set in a complete lattice. If this process is applied to an O-group, the result in general is not a group. However, if the O-group can be O-embedded in a complete ℓ-group, then the embedding can be achieved by the Dedekind cut process. This process is described in detail in Fuchs' book [11].

The embedding in a complete ℓ-group has been studied by several authors including Clifford [5], Conrad and McAlister [7], Krull [19], and Lorenzen [21]. The following proof employs the notion of v-ideal. Recall that if D is a Bezout domain with archimedean group of divisibility, then D is completely integrally closed [13, p. 614].

THEOREM 4.5. *If G is an archimedean ℓ-group, then G can be embedded in a complete ℓ-group.*

Proof. Let D be a Bezout domain with G as its group of divisibility. Since G is archimedean, D is completely integrally closed, and $\mathcal{V}(D)$ is a complete ℓ-group. Clearly $G(D) = G$ can be embedded in $\mathcal{V}(D)$.

Observe that in the sense of Theorem 1.1 of [7], $\mathcal{V}(D)$ is the completion of $G(D)$.

If G is a lattice-ordered group and $g \in G$, $g \neq 0$, then a *value* of g is a convex ℓ-subgroup M_g such that for any convex ℓ-subgroup $H \supset M_g$, $g \in H$. A Zorn's lemma argument shows the existence of a value for any non-zero $g \in G$. It is well known that G/M_g is totally ordered; that is, M_g is a prime subgroup of G [7, p. 188]. I will give a proof using Theorem 2.1.

PROPOSITION 4.6. *Suppose that G is a lattice-ordered group and that g is a non-zero element of G. If M_g is a value of G, then G/M_g is totally ordered.*

Proof. Without loss of generality assume $g > 0$. Let D be a Bezout domain with group of divisibility G. We show M_g corresponds to a prime ideal of D. If v is the canonical semi-valuation, $S = v^{-1}((M_g)_+)$ is a saturated multiplicative system since M_g is a convex directed subgroup of G [23]. If $x \in D$ is such that $v(x) = g$, then $x \notin S$. Let P be a prime ideal of D containing x such that $P \cap S = \emptyset$. Thus $D \setminus P$ is a saturated multiplicative system containing S. Hence, $v(D \setminus P)$ generates a convex ℓ-subgroup H of G containing M_g. Furthermore, $g \notin H$. Hence, $H = M_g$, $S = D \setminus P$, and $G/_{M_g}$ is the group of divisibility of the valuation ring D_P [23].

Finally, I apply Theorem 2.1 to obtain a ring theoretic proof of a well known result about realizations of an ℓ-group by subgroups of reals. First, observe the following proposition.

PROPOSITION 4.7. _If D is a Bezout domain and if $\{V_\alpha\}$ is a collection of valuation overrings of D such that each non-zero non-unit x of D is contained in the center of some V_α, then $xD = \cap_\alpha (xV_\alpha \cap D)$ and $D = \cap_\alpha V_\alpha$._

The proof is similar to that of [13, p. 42] needing only the additional observation that $[zD:xD]$ is finitely generated [13, p. 301].

An archimedean totally ordered group can be embedded as a subgroup of the reals. This fact led Clifford [5] to conjecture that an archimedean ℓ-group has a realization by subgroups of R. Similarly, Krull [19] conjectured that a completely integrally closed domain is an intersection of rank one valuation rings. Nakayama gave a counterexample to both conjectures in [25]. Realizations by subgroups of R are obtained by applying the following corollary to Proposition 4.7.

COROLLARY 4.8. _If D is a Bezout domain such that each non-zero non-unit of D is contained in a prime ideal of height one, then the group of divisibility can be embedded as an ℓ-subgroup of a cardinal product of subgroups of the reals._

Proof. By Proposition 4.7, D is an intersection of valuation rings D_P where P runs through all prime ideals of D of height one. Apply Theorem 2.2. Since P is of height one, D_P is a rank one valuation ring, and $G(D_P)$ is a subgroup of R.

The next theorem in its present form appeared in a paper by Conrad, Harvey, and Holland [8, p. 164], but it was originally proved for vector lattices by Nakayama [26]. I add part 3 for completeness.

THEOREM 4.9. (Nakayama, Conrad, Harvey, Holland) _Suppose_ G _is a lattice-ordered abelian group. The following are equivalent._

1) G _is_ 0-_isomorphic to a sublattice of the cardinal product of copies of the additive group of real numbers._

2) _For each non-zero_ $g \in G$, _there is a value_ M_g _of_ g _such that_ $G/_{M_g}$ _is_ ℓ-_isomorphic to a subgroup of the real numbers._

3) G _is the group of divisibility of a Bezout domain_ D _such that each non-zero non-unit_ x _of_ D _is contained in a prime ideal of height one._

Proof. That 3) implies 1) is the content of Corollary 4.8. Clearly 1) implies 2). Finally, assume 2) holds. By Theorem 2.1, there is a Bezout domain D with G as its group of divisibility. Suppose x is a non-zero non-unit of D. If v is the canonical semi-valuation onto G, let $g = v(x)$. Let M_g be a value of g such that $G/_{M_g}$ is contained in the reals. By [23], M_g corresponds to a saturated multiplicative system S of D. Since $G/_{M_g}$ is totally ordered, S is the complement of a prime ideal P of D. Further $G(D_S) = G/_{M_g}$ and $G/_{M_g}$ a subgroup of the reals imply that D_S is a rank one valuation ring and that P is a prime ideal of height one.

REFERENCES

1. Arnold, J. On the ideal theory of the Kronecker function ring and the domain D(X), _Canad. J._ 21, 558-563 (1969).

2. Birkhoff, G. Lattice-ordered groups, _Annals of Math._ 43, 298-331 (1942).

3. Bourbaki, N. _Algebra Commutative_, Chapitre 7, Herman, Paris, 1965.

4. Brewer, J. and Mott, J. L. On integral domain of finite character, _J. Reine Angew. Math._ 241, 34-41 (1970).

5. Clifford, A. H. Partially ordered abelian groups, _Ann. of Math._ 41, 465-473 (1940).

6. Cohen, I. S. and Kaplansky, I. Rings with a finite number of primes, I. Trans. Amer. Math. Soc. 60 (1946), 468-477.

7. Conrad, P. F. and McAlister, D. The completion of a lattice-ordered group, J. Australian Math. Soc. 9, 182-208 (1969).

8. Conrad, P., Harvey, J., and Holland, C. The Hahn embedding theorem for abelian lattice-ordered groups, Trans. Amer. Math. Soc. 108, 143-169 (1963).

9. Dedekind, R. Ueber Zerlegungen von Zahlen durch ihre grossten gemeinsamen Teiler, Gesammelte Math. Werke, Vol. II, Chelsea, New York, 103-147 (1969).

10. Dieudonné, J. Sur la Théorie de la Divisibilité, Bull. Soc. Math. France 49, 1-12 (1941).

11. Fuchs, L. Partially ordered algebraic systems, Pergamon Press, New York, 1963.

12. ________ The generalization of the valuation theory, Duke Math. J. 18, 19-26 (1951).

13. Gilmer, R. W. Multiplicative ideal theory, Queens' Papers, Lecture Notes No. 12, Queen's University, Kingston, Ontario, 1968.

14. Heinzer, W. Some remarks on complete integral closure, J. Australian Math. 9, 310-314 (1969).

15. ________ J-Noetherian integral domains with 1 in the stable range, Proc. Amer. Math. Soc. 19, 1369-1372 (1968).

16. Jaffard, P. Contribution à la théorie des groupes ordonnés, J. Math. Pures Appl. 32, 203-280 (1953).

17. ________ Extension des groupes réticulés et applications, Publ. Sci. Univ. Alger. Ser. A-1, 197-222 (1954).

18. ________ Un contre-exemple concernant les groupes de divisibilité, C. R. Acad. Sci. Paris 243, 1264-1268 (1956).

19. Krull, W. Allgemeine Bewertungetheorie, J. Reine Angew. Math. 167, 160-196 (1931).

20. ________ Beiträge zur arithmetik kommutativer Integritätsbereiche; I, II, Math. Z. 41, 545-577; 665-679 (1936).

21. Lorenzen, P. Abstrakte Begrundung der multiplicativen Idealtheorie, Math. Z. 45, 533-553 (1939).

22. MacNeille, H. Partially ordered sets, Trans. Amer. Math. Soc. 42, 416-460 (1937).

23. Mott, J. L. The convex directed subgroups of a group of divisibility, submitted for publication.

24. Nagata, M. Local rings, Interscience New York, 1962.

25. Nakayama, T. On Krull's conjecture concerning completely integrally closed integrity domains, I, II, Proc. Imp. Acad. Tokyo 18, 185-187; 233-236 (1942); III, Proc. Japan Acad. 22, 249-250 (1946).

26. _______ Note on lattice-ordered groups, Proc. Imp. Acad. Tokyo 18, 1-4 (1942).

27. Ohm, J. Some counterexamples related to integral closure in D[[X]], Trans. Amer. Math. Soc. 122, 321-333 (1966).

28. _______ Semi-valuation and groups of divisibility, Canad. J. Math. 21, 576-591 (1969).

29. Prüfer, H. Untersuchungen über die Teilbarkeitseigenschaften in Körpern, J. Reine Angew. Math. 168, 1-36 (1932).

30. Ribenboim, P. Conjonction d'ordres dans les groupes abelian ordonnés, An. Acad. Brasil. Ci. 29, 201-224 (1957).

31. _______ Un théorème de realisation de groupes réticulés, Pacific J. Math. 10, 305-308 (1960).

32. Sheldon, P. A counter example to a conjecture of Heinzer, submitted for publication.

33. Ward, W. Residuated lattices, Duke Math. J. 6, 641-651 (1940).

34. van der Waerden, B. L. Modern Algebra, Vol. II, 2nd English edition, Ungar, New York, 1950.

HOMOLOGICAL DIMENSION AND EULER MAPS*

Jack Ohm, Purdue University
and Louisiana State University

1. Introduction. If $\mathcal{M}$ is a collection of R-modules, an Euler map for $\mathcal{M}$ is a map f of $\mathcal{M}$ into an abelian group G which is multiplicative on short exact sequences from $\mathcal{M}$. For example, if $\mathcal{M}$ is the set of R-modules M which have a finite free resolution, then f(M) can be taken to be the alternating sum of the ranks of the free modules in a resolution for M; here the group G is the additive group of integers. Another example is that of MacRae [M], who has defined an Euler map from the set of torsion R-modules of finite projective dimension into the multiplicative group of invertible fractional ideals of R.

I shall give here a unified approach to these examples. My only claim to originality is in providing what I feel is a proper setting for the exposition; otherwise I have followed the paper [M] of MacRae and Kaplansky's treatment in [K]. (Compare also Bass [B] and Swan-Evans [SE]). Roughly, the idea is to begin with a set of R-modules $\mathcal{M}$ and a subset $\mathcal{O}$ of $\mathcal{M}$, both of which are closed under direct sum. If then an Euler map can be defined on $\mathcal{O}$, under certain conditions it can be uniquely extended to an Euler map on the set of elements of $\mathcal{M}$ having finite $\mathcal{O}$-resolution. The $\mathcal{M}$ and $\mathcal{O}$ for the first of the above examples can be taken to be the set of all finitely generated R-modules and the set of all finitely generated free R-modules respectively, while the $\mathcal{M}$ and $\mathcal{O}$ which give the Euler map of MacRae are more complicated and will be discussed in §3.

* This research was supported by NSF Grant GP-29104A.

2. Dimension. Let $\mathscr{M}$ be an abelian semigroup with respect to an operation $+$. We need the notion of an exact sequence, and one possibility would be to work in an abelian category. However, to avoid extraneous considerations, we have chosen instead to write down the properties required for our theorems and then to define an exact sequence to be something having these properties. In particular, morphisms never enter the picture; only the existence of certain exact sequences matters. Thus, we shall suppose there is given a collection $\mathscr{E}$ of sequences of the form $(M_n,\ldots,M_2, M_1)$, $n \geq 3$, $M_i \in \mathscr{M}$, which satisfies the following seven axioms:

$\mathscr{E}$1. $(M_n,\ldots,M_1) \in \mathscr{E} \Rightarrow (M_{i+j},\ldots,M_{i+1},M_i) \in \mathscr{E}$ for every $i = 1,\ldots,n-2$ and j such that $2 \leq j \leq n-i$.

$\mathscr{E}$2. For any $M \in \mathscr{M}$ and any i such that $1 \leq i < n$, $(M_n,\ldots,M_1) \in \mathscr{E} \Rightarrow (M_n,\ldots,M_{i+2}, M_{i+1} + M, M_i + M, M_{i-1},\ldots,M_1) \in \mathscr{E}$ (i.e. M can be added to two successive terms of any element of $\mathscr{E}$).

$\mathscr{E}$3. For every $M \in \mathscr{M}$, $(0,0,M,M,0,0) \in \mathscr{E}$.

$\mathscr{E}$4. $(0,M',M,0) \in \mathscr{E} \Rightarrow M' = M$.

$\mathscr{E}$5. (Factoring through images). For every i such that $1 \leq i < n$, $(M_n,\ldots,M_1) \in \mathscr{E} \Leftrightarrow$ there exists $K_i \in \mathscr{M}$ such that $(M_n,\ldots,M_{i+1}, K_i, 0) \in \mathscr{E}$ and $(0, K_i, M_i,\ldots,M_1) \in \mathscr{E}$.

$\mathscr{E}$6. $(0,A,B,C,0) \in \mathscr{E}$ and $(0,K,M,B,0) \in \mathscr{E} \Rightarrow$ there exists $M' \in \mathscr{M}$ such that $(0,M',M,C,0) \in \mathscr{E}$ and $(0,K,M',A,0) \in \mathscr{E}$.

$\mathscr{E}$7. (Existence of pullbacks). $(0,A,B,C,0) \in \mathscr{E}$ and $(0,A',B',C,0) \in \mathscr{E} \Rightarrow$ there exists $M \in \mathscr{M}$ such that $(0,A,M,B',0) \in \mathscr{E}$ and $(0,A',M,B,0) \in \mathscr{E}$.

An element of the set $\mathscr{E}$ will be called an exact sequence. If $(M_n,\ldots,M_1)$ is exact, we shall call a K_i given by $\mathscr{E}$5 an image for (M_{i+1},M_i) and shall write $K_i = \text{im}\,(M_{i+1},M_i)$.

2.1 <u>Consequences of the axioms</u>. i) $(0,M,0) \in \mathcal{S} \Rightarrow M = 0$, by $\mathcal{S}1$, $\mathcal{S}3$, $\mathcal{S}4$, and $\mathcal{S}5$.

ii) For any $M', M \in \mathcal{M}$, $(0,M',M' + M,M,0) \in \mathcal{S}$, by $\mathcal{S}1$, $\mathcal{S}2$, and $\mathcal{S}3$.

Note that if $\mathcal{M}$ is the semigroup of isomorphism classes of R-modules (for a fixed ring R) under direct sum, the above axioms are satisfied by the exact sequences in the usual sense. For $\mathcal{S}6$, if $\phi: M \to B$, take $M' = \phi^{-1}(A)$; and for $\mathcal{S}7$, if $\psi: B \to C$, $\psi': B' \to C$, take $M = \{(b,b') \in B \oplus B' \mid \psi(b) = \psi'(b')\}$. The appropriate diagrams for $\mathcal{S}6$ and $\mathcal{S}7$ are the following:

$\mathcal{S}6.$

$$\begin{array}{ccccccccc} & & 0 & & 0 & & & & \\ & & \downarrow & & \downarrow & & & & \\ & & K & & K & & & & \\ & & \downarrow & & \downarrow & & & & \\ 0 & \dashrightarrow & M' & \dashrightarrow & M & \dashrightarrow & C & \dashrightarrow & 0 \\ & & \downarrow & & \downarrow & & & & \\ 0 & \to & A & \to & B & \to & C & \to & 0 \\ & & \downarrow & & \downarrow & & & & \\ & & 0 & & 0 & & & & \end{array}$$

$\mathcal{S}7.$

$$\begin{array}{ccccccccc} & & & & 0 & & 0 & & \\ & & & & \downarrow & & \downarrow & & \\ & & & & A' & & A' & & \\ & & & & \downarrow & & \downarrow & & \\ 0 & \dashrightarrow & A & \dashrightarrow & M & \dashrightarrow & B' & \dashrightarrow & 0 \\ & & & & \downarrow & & \downarrow & & \\ 0 & \to & A & \to & B & \to & C & \to & 0 \\ & & & & \downarrow & & \downarrow & & \\ & & & & 0 & & 0 & & \end{array}$$

Given an exact sequence $(0,A,B,C,0)$, $A,B,C \in \mathcal{M}$, an element $P \in \mathcal{M}$ will be called <u>projective relative to $(0,A,B,C,0)$</u> if for any $K,K',L' \in \mathcal{M}$ such that $(0,K',L',A,0)$ and $(0,K,P,C,0)$ are exact, there exists $M \in \mathcal{M}$ such that $(0,K',M,K,0)$ and $(0,M,L' + P,B,0)$ are exact. One should have in mind the following diagram:

$$\begin{array}{ccccccccc} & & 0 & & 0 & & 0 & & \\ & & \downarrow & & \downarrow & & \downarrow & & \\ 0 & \dashrightarrow & K' & \dashrightarrow & M & \dashrightarrow & K & \dashrightarrow & 0 \\ & & \downarrow & & \downarrow & & \downarrow & & \\ & & L' & & L'+P & & P & & \\ & & \downarrow & & \downarrow & & \downarrow & & \\ 0 & \to & A & \to & B & \to & C & \to & 0 \\ & & \downarrow & & \downarrow & & \downarrow & & \\ & & 0 & & 0 & & 0 & & \end{array}$$

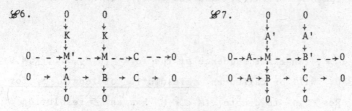

Note that projectives split, in the sense that $(0,A,B,P,0)$ exact and P projective relative to $(0,A,B,P,0)$ implies $B = A + P$. One sees this by taking $K = K' = 0$ and $L' = A$ in the above. Then $M = 0$ by 2.1, and hence $B = A + P$ by $\mathcal{S}4$. If P is projective relative to every short exact sequence $(0,A,B,C,0)$, $A,B,C \in \mathcal{M}$, then we shall simply call P projective.

Let $\mathcal{O}$ be a subsemigroup of $\mathcal{M}$. We define an equivalence relation on $\mathcal{M}$ as follows:

For any $M_1, M_2 \in \mathcal{M}$, $M_1 \sim M_2$ if there exist $P_1, P_2 \in \mathcal{O}$ such that $M_1 + P_1 = M_2 + P_2$.

The equivalence class $\mathcal{O}^*$ of $\mathcal{O}$ is then $\{M \in \mathcal{M} \mid$ there exists $P \in \mathcal{O}$ such that $M + P \in \mathcal{O}\}$, and one easily verifies that $\mathcal{O}^{**} = \mathcal{O}^*$.

If $M \in \mathcal{M}$, an exact sequence of the form $(0,P_n,\ldots,P_0,M,0)$, $P_i \in \mathcal{O}$, $n \geq 0$, will be called an $\mathcal{O}$-resolution of length n for M. We use $\mathrm{Res}(\mathcal{O},\mathcal{M})$ to denote $\{M \in \mathcal{M} \mid M$ has an $\mathcal{O}$-resolution of finite length$\}$. If $M \neq 0 \in \mathrm{Res}(\mathcal{O},\mathcal{M})$, we define the $\mathcal{O}$-dimension of M, $d_{\mathcal{O}}(M)$, to be the least n such that M has an $\mathcal{O}$-resolution of length n; we define $d_{\mathcal{O}}(0) = -1$.

2.2 Proposition. Let $(K,P_n^*,\ldots,P_0^*,M,0)$ be exact with $P_i^* \in \mathcal{O}^*$ and $n > 0$. Then there exist $P_i \in \mathcal{O}$ such that $P_i^* \sim P_i, i = 0,\ldots,n$ and such that $(K,P_n,\ldots,P_0,M,0)$ is exact.

Proof. Since $P_0^*, P_1^* \in \mathcal{O}^*$, there exists $Q \in \mathcal{O}$ such that $P_0 = P_0^* + Q$ and $P_1 = P_1^* + Q$ are in $\mathcal{O}$. By $\mathcal{S}2$, $(K,P_n^*,\ldots,P_2^*,P_1,P_0,M,0)$ is exact. If $n = 1$, this is the required sequence. If $n > 1$, factor through $K_0 = \mathrm{im}(P_1,P_0)$ via $\mathcal{S}5$ and apply an induction hypothesis.

2.3 Corollary. $\mathrm{Res}(\mathcal{O},\mathcal{M}) = \mathrm{Res}(\mathcal{O}^*,\mathcal{M})$. If $M \in \mathrm{Res}(\mathcal{O},\mathcal{M})$, then $d_{\mathcal{O}}(M) = d_{\mathcal{O}^*}(M)$ if $d_{\mathcal{O}^*}(M) \neq 0$, and $0 \le d_{\mathcal{O}}(M) \le 1$ if $d_{\mathcal{O}^*}(M) = 0$.

2.4 Proposition. Suppose the elements of $\mathcal{O}$ are projective and $(0,K,P_{n-1},\dots,P_0,M,0)$ and $(0,K',P'_{n-1},\dots,P_0',M',0)$ are exact with $P_i,P_i' \in \mathcal{O}^*$ and $n \ge 0$. If $M \sim M'$, then $K \sim K'$.

Proof. Proceed by induction on n. $n = 0$: $(0,K,M,0)$ and $(0,K',M',0)$ exact imply $K = M \sim M' = K'$ by $\mathcal{C}4$.
$n = 1$: Since $M \sim M'$, there exist $P,P' \in \mathcal{O}$ such that $M + P = M' + P'$. Therefore by adding on a summand via $\mathcal{C}2$, it suffices to prove the proposition when $M = M'$. Similarly, we may assume $P_0,P_0' \in \mathcal{O}$. It follows from the existence of pullbacks and the splitting of projectives (as with Schanuel's lemma) that $K + P_0' = K' + P_0$. Hence $K \sim K'$.

For the induction step, factor through $K_0 = \mathrm{im}(P_1,P_0)$ and $K_0' = \mathrm{im}(P_1',P_0')$. $K_0 \sim K_0'$ by the $n = 1$ case, and then $K \sim K'$ by the induction hypothesis.

2.5 Corollary. Suppose the elements of $\mathcal{O}$ are projective, and let $M \neq 0$ be an element of $\mathcal{M}$ having an $\mathcal{O}^*$ resolution $(0,P_n^*,\dots,P_0^*,M,0)$. Let $K_0 = \mathrm{im}(P_0^*,M)$, and let $K_i = \mathrm{im}(P_i^*,P_{i-1}^*)$, $1 \le i \le n$. Then $d_{\mathcal{O}^*}(M) = \min\{i \mid K_i \in \mathcal{O}^*\}$.

The above corollary shows that when the elements of $\mathcal{O}$ are projective, one need only examine a given $\mathcal{O}^*$-resolution for M to compute $d_{\mathcal{O}^*}(M)$, rather than check all $\mathcal{O}^*$-resolutions and select a minimal one. For this reason, we shall henceforth in the remainder of §2 use only $\mathcal{O}^*$ to compute dimension, which in turn allows us to abbreviate our notation to <u>d(M)</u> for $d_{\mathcal{O}^*}(M)$ and

Res for Res($\mathcal{O}^*$,$\mathcal{M}$) (≈ Res($\mathcal{O}$,$\mathcal{M}$)). We shall also assume for the remainder of §2 that the elements of $\mathcal{O}$ are projective.

2.6 Corollary. If $M \neq 0$ and $M' \neq 0$ are in Res and $M \sim M'$, then $d(M) = d(M')$.

Proof. Apply 2.4 and 2.5.

We shall show in 2.8 that the hypothesis that $M' \in$ Res can be deleted from 2.6.

2.7 Lemma. i) Suppose $M,K \in$ Res, $P \in \mathcal{O}^*$, and $(0,K,P,M,0)$ is exact. If $d(M) > 0$, then $d(M) = 1 + d(K)$; while if $d(M) = 0$, then $d(K) \leq 0$.

ii) If $M \neq 0$ is in Res, then there exist $P \in \mathcal{O}^*$ and $K \in$ Res such that $(0,K,P,M,0)$ is exact and $d(M) = 1 + d(K)$.

Proof. (i) Suppose $d(M) = n > 0$, and let $(0,P_n,\ldots,P_0,M,0)$ be a minimal $\mathcal{O}^*$-resolution for M. Factoring through $K_1 = \mathrm{im}(P_1,P_0)$, we have $(0,P_n,\ldots,P_1,K_1,0)$ is a minimal $\mathcal{O}^*$-resolution for K_1. Also, $n > 0$ implies $K \neq 0$, $K_1 \neq 0$. Since $K \sim K_1$ by 2.4, it follows from 2.6 that $d(K) = n - 1$. Suppose now $d(M) = 0$. By adding a summand to P and M via $\mathcal{S}$2, we may assume $P,M \in \mathcal{O}$. Then by splitting of projectives, $P = K + M$; and hence $K \in \mathcal{O}^*$ and $d(K) \leq 0$.
(ii) Let $(0,P_n,\ldots,P_0,M,0)$ be a minimal $\mathcal{O}^*$-resolution for M, and factor through $K = \mathrm{im}(P_1,P_0)$. Then $K \in$ Res and $(0,K,P_0,M,0)$ is the required sequence.

2.8 Lemma. If $M \in$ Res and $M' \sim M$, then $M' \in$ Res.

Proof. It suffices to see that for $P \in \mathcal{O}$, $M \in$ Res if and only if $M + P \in$ Res. That $M \in$ Res implies $M + P \in$ Res is an immediate consequence of $\mathcal{S}2$. Conversely, suppose $M + P \in$ Res. $(0,P,M + P,M,0)$ is exact by 2.1. Moreover, $M + P = 0$ implies $M = 0$ by 2.1, so we may assume $M + P \neq 0$. By 2.7, there exist $P' \in \mathcal{O}^*$ and $K \in$ Res such that $(0,K,P',M + P,0)$ is exact. By $\mathcal{S}6$, there exists $L \in \mathcal{M}$ such that $(0,K,L,P,0)$ and $(0,L,P',M,0)$ are exact. Then by splitting of projectives, P projective implies $L = K + P$; and $K + P \in$ Res since $K \in$ Res. But then $(0,L,P',M,0)$ is exact and $L \in$ Res imply $M \in$ Res.

2.9 Theorem (dimension theorem). Let A,B,C be elements of $\mathcal{M}$ such that $(0,A,B,C,0)$ is exact. If any two of A,B,C are in Res, then the third is also. When this occurs, $d(B) \leq \max\{d(A),d(C)\}$; and if the inequality holds, then $d(C) = d(A) + 1$.

Proof. Case (i): Assume A and C are in Res. We proceed by induction on $d(C)$. If $d(C) = -1$, then $C = 0$ and $A = B$ and the theorem is immediate. Assume then that $d(C) = n > -1$. Then $C \neq 0$, and hence also $B \neq 0$. If $A = 0$, then $B = C$ and the theorem follows; so we may also assume $A \neq 0$. By 2.7(ii), there exist $P,P' \in \mathcal{O}^*$ and $K,K' \in$ Res such that $(0,K',P',A,0)$ and $(0,K,P,C,0)$ are exact and such that $d(A) = 1 + d(K')$ and $d(C) = 1 + d(K)$. Since an $\mathcal{O}$-summand does not alter dimension or the property of being in Res, by 2.6 and 2.8, by adding on an element of $\mathcal{O}$ to B,C, and P via $\mathcal{S}2$, we may assume $P \in \mathcal{O}$ and hence is projective. Then there exists $L \in \mathcal{M}$ such that $(0,K',L,K,0)$ and $(0,L,P' + P,B,0)$ are exact. By the induction hypothesis, $L \in$ Res, and hence $B \in$ Res, and also

$d(L) \leq \max\{d(K'),d(K)\}$ and * holds implies $d(K) = d(K') + 1$. If $d(B) > 0$, then by 2.7(i), $d(B) \doteq d(L) + 1$; and then the dimensions of K',L,K are all one less than those of A,B,C respectively, and the assertion for K',L,K thus carries down to A,B,C. If $d(B) = 0$, then $B \in \mathcal{O}^*$ and the theorem follows from 2.7(i) applied to $(0,A,B,C,0)$. This concludes case (i).

It only remains to establish that (Case (ii)) $A,B \in$ Res implies $C \in$ Res and (Case (iii)) $B,C \in$ Res implies $A \in$ Res. We can as before dispose of the case $B = 0$, so assume $B \in$ Res and $B \neq 0$. Then there exist $Q \in \mathcal{O}^*$ and $M \in$ Res such that $(0,M,Q,B,0)$ is exact and $d(B) = 1 + d(M)$; and hence by $\mathcal{S}$6 there exists $N \in \mathcal{M}$ such that $(0,N,Q,C,0)$ and $(0,M,N,A,0)$ are exact. Case (ii): M and A are in Res implies $N \in$ Res by case (i), and then $C \in$ Res since $(0,N,Q,C,0)$ is exact. Case (iii): $C \in$ Res implies there exist $N' \in$ Res and $Q' \in \mathcal{O}^*$ such that $(0,N',Q',C,0)$ is exact. By 2.4, $N \sim N'$; and hence $N \in$ Res by 2.8. Then M and $N \in$ Res implies $A \in$ Res by case (ii).

3. Euler maps. Let $\mathcal{J}$ be a subset of $\mathcal{M}$. An Euler map for $\mathcal{J}$ is a map f of $\mathcal{J}$ into a (multiplicative) abelian group G such that for any short exact sequence $(0,A,B,C,0)$ with $A,B,C \in \mathcal{J}$, $f(A)f(C) = f(B)$. A simple induction shows that then for any exact sequence $(0,M_n,\ldots,M_1,0)$, $M_i \in \mathcal{J}, n \geq 3$, one has $f(M_1) = f(M_2)f(M_3)^{-1}f(M_4)\ldots$.

3.1 Let $\mathcal{O}_i \subset \mathcal{O}_{i+1}$ be subsets of $\mathcal{M}$ such that

a) If $(0,A,B,C,0)$ is exact and A,C are in $\mathcal{O}_i$, then $B \in \mathcal{O}_i$.

b) For any $M \in \mathcal{O}_{i+1}$, there exist $P,K \in \mathcal{O}_i$ such that

$(0,K,P,M,0)$ is exact; moreover, given any exact sequence $(0,A,B,M,0)$ with $A,B \in \mathcal{O}_{i+1}$, this P can be chosen to be projective relative to $(0,A,B,M,0)$.

Theorem. If there exists an Euler map f of $\mathcal{O}_i$ into an abelian group G, then f extends uniquely to an Euler map of $\mathcal{O}_{i+1}$ into G.

Proof. If $M \in \mathcal{O}_{i+1} \setminus \mathcal{O}_i$, then by (b) there exist $P,K \in \mathcal{O}_i$ such that $(0,K,P,M,0)$ is exact. Define (necessarily) $f(M) = f(P)f(K)^{-1}$.

If P',K' are also elements of $\mathcal{O}_i$ such that $(0,K',P',M,0)$ is exact, then by the existence of pullbacks, there exists $W \in \mathcal{M}$ such that $(0,K,W,P',0)$ and $(0,K',W,P,0)$ are exact. Then $K,P' \in \mathcal{O}_i$ implies $W \in \mathcal{O}_i$ by (a). Therefore $f(K)f(P') = f(W) = f(K')f(P)$ since f is an Euler map on $\mathcal{O}_i$. Thus, $f(P)f(K)^{-1} = f(P')f(K')^{-1}$ and $f(M)$ is independent of the choice of $P,K \in \mathcal{O}_i$.

Now let us prove that for any $A,B,C \in \mathcal{O}_{i+1}$ such that $(0,A,B,C,0)$ is exact, $f(A)f(C) = f(B)$. By (b), there exist $P',P,K',K \in \mathcal{O}_i$ such that $(0,K',P',A,0)$ and $(0,K,P,C,0)$ are exact and such that P is projective relative to $(0,A,B,C,0)$. By the projectivity of P, there exists $L \in \mathcal{M}$ such that $(0,K',L,K,0)$ and $(0,L,P'+P,B,0)$ are exact. Then $K,K' \in \mathcal{O}_i$ implies $L \in \mathcal{O}_i$ by (a), and hence $f(K)f(K') = f(L)$. Similarly, $f(P'+P) = f(P')f(P)$. We have already checked that f is multiplicative on exact sequences of the form $(0,K,P,M,0)$ with $K,P \in \mathcal{O}_i$ and $M \in \mathcal{O}_{i+1}$; so it follows that $f(A) = f(P')f(K')^{-1}, f(B) = f(P'+P)f(L)^{-1}$, and $f(C) = f(P)f(K)^{-1}$. Putting these equalities together, we get the desired result.

3.2 Definition. Let $\mathscr{S} \subset \mathscr{S}'$ be subsets of $\mathscr{M}$. $\mathscr{S}$ will be said to contain sufficient projectives relative to $\mathscr{S}'$ if for any $M \in \mathscr{S}'$ there exist $P \in \mathscr{S}$ and $K \in \mathscr{M}$ such that $(0,K,P,M,0)$ is exact and if for any given exact sequence $(0,A,B,M,0)$ with $A,B \in \mathscr{S}'$, this P can be chosen to be projective relative to $(0,A,B,M,0)$.

For example, if the elements of $\mathscr{O}$ are projective, then $\mathscr{O}$ contains sufficient projectives relative to $\operatorname{Res}(\mathscr{O},\mathscr{M})$. Note that if $\mathscr{S}$ contains sufficient projectives relative to $\mathscr{S}'$, then the same is true for any $\mathscr{S}''$ such that $\mathscr{S} \subset \mathscr{S}'' \subset \mathscr{S}'$.

We retain here the notation of §2: $\mathscr{O}$ denotes a subsemigroup of $\mathscr{M}$ consisting of projectives, and $d(\ )$ denotes $\mathscr{O}^*$-dimension. Let $\mathscr{T}$ be a subset of $\mathscr{M}$ such that if $(0,A,B,C,0)$ is exact and any two of A,B,C are in $\mathscr{T}$, then the third is also; let $\operatorname{Res}(\mathscr{O},\mathscr{T}) = \operatorname{Res}(\mathscr{O},\mathscr{M}) \cap \mathscr{T}$; and let $\operatorname{Res}_i(\mathscr{O},\mathscr{T}) = \{M \in \operatorname{Res}(\mathscr{O},\mathscr{T}) \mid d(M) \le i\}$.

3.3 Theorem. If for some n $\operatorname{Res}_n(\mathscr{O},\mathscr{T})$ contains sufficient projectives relative to $\operatorname{Res}(\mathscr{O},\mathscr{T})$ and there exists an Euler map f of $\operatorname{Res}_n(\mathscr{O},\mathscr{T})$ into an abelian group G, then f extends uniquely to an Euler map of $\operatorname{Res}(\mathscr{O},\mathscr{T})$ into G.

Proof. Since $\operatorname{Res}(\mathscr{O},\mathscr{T}) = \cup_{i\ge n}\operatorname{Res}_i(\mathscr{O},\mathscr{T})$, it suffices to show that an Euler map of $\mathscr{T}_i = \operatorname{Res}_i(\mathscr{O},\mathscr{T}), i \ge n$, extends uniquely to an Euler map of $\mathscr{T}_{i+1} = \operatorname{Res}_{i+1}(\mathscr{O},\mathscr{T})$. This will follow from 3.1 provided we verify (a) and (b) of 3.1. That (a) is satisfied is a consequence of the dimension theorem. Since $\mathscr{T}_i$ contains sufficient projectives relative to $\operatorname{Res}(\mathscr{O},\mathscr{T})$, for any $M \in \mathscr{T}_{i+1}$, there exists $P \in \mathscr{T}_i$ and $K \in \mathscr{M}$ which satisfy the requirements of (b), except possibly for the condition that $K \in \mathscr{T}_i$, and this follows again from the dimension theorem.

3.4 Application. Fix a commutative ring with identity R. Let $\mathcal{M}$ be the semigroup of isomorphism classes of R-modules under direct sum, let $\mathcal{O}_p$ be the subset corresponding to the finitely generated projective modules, and let $\mathcal{T}$ be the subset corresponding to the torsion modules. (An R-module M is called a torsion R-module if there exists a regular element $r \in R$ such that $rM = 0$.) Take the set $\mathcal{S}$ of exact sequences to be the usual exact sequences, i.e. if $\tilde{M}$ denotes the isomorphism class of the R-module M, $(\tilde{M}_n, \ldots, \tilde{M}_1)$ is exact if and only if there exist homomorphisms $\rightarrow$ such that $M_n \rightarrow \ldots \rightarrow M_1$ is exact.

Lemma. $\mathrm{Res}_1(\mathcal{O}_p, \mathcal{T})$ contains sufficient projectives relative to $\mathrm{Res}(\mathcal{O}_p, \mathcal{T})$.

Proof. Suppose $\tilde{M} \in \mathrm{Res}(\mathcal{O}_p, \mathcal{T})$. Then there exist $\tilde{P} \in \mathcal{O}_p$ and $\tilde{K} \in \mathcal{M}$ such that $0 \rightarrow K \rightarrow P \rightarrow M \rightarrow 0$ is exact. Therefore if r is any regular element of R such that $rM = 0$, then

$$0 \rightarrow K/rP \rightarrow P/rP \rightarrow M \rightarrow 0$$

is also exact, and $(P/rP)^{\sim} \in \mathrm{Res}_1(\mathcal{O}_p, \mathcal{T})$. Suppose now we are given an exact sequence $0 \rightarrow A \rightarrow B \rightarrow M \rightarrow 0$, with $\tilde{A}, \tilde{B} \in \mathrm{Res}(\mathcal{O}_p, \mathcal{T})$. Since $\tilde{B} \in \mathcal{T}$, there exists a regular $r \in R$ such that $rB = 0$ and hence such that $rM = 0$. We shall show that P/rP is projective relative to $0 \rightarrow A \rightarrow B \rightarrow M \rightarrow 0$. Thus, suppose there is given a homomorphism $P/rP \rightarrow M$. Consider the following diagram:

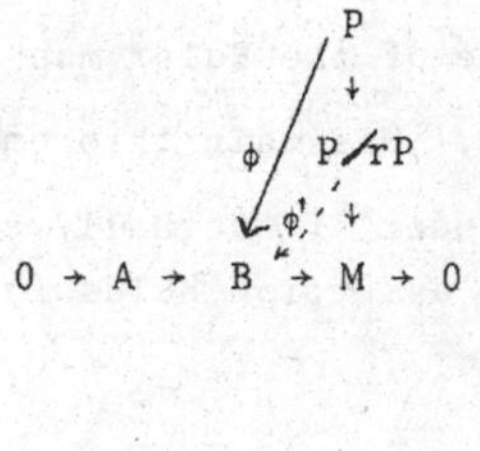

The existence of a ϕ making the outer triangle commutative results from P being projective, and then $rP \subset$ kernel ϕ implies the existence of a ϕ' making the diagram commutative. Now given any exact sequences $0\to K'\to L'\to A\to 0$ and $0\to K\to P/rP\to M\to 0$, we can find via the above ϕ' homomorphisms which make the following diagram commutative with exact rows and columns:

$$\begin{array}{ccccccccc} & & 0 & & & & 0 & & \\ & & \downarrow & & & & \downarrow & & \\ & & K' & & & & K & & \\ & & \downarrow & & & & \downarrow & & \\ 0 & \to & L' & \to & L'\oplus(P/rP) & \to & P/rP & \to & 0 \\ & & \downarrow & & \downarrow & & \downarrow & & \\ 0 & \to & A & \to & B & \to & M & \to & 0 \\ & & \downarrow & & \downarrow & & \downarrow & & \\ & & 0 & & 0 & & 0 & & \end{array}$$

If N denotes the kernel of $L' \oplus P/rP \to B$, then by the 9-lemma [Mi, p.20], this is the element required to prove P/rP is projective relative to $0\to A\to B\to M\to 0$. Q.E.D.

MacRae [M, §2] has shown that the map which assigns to each element of $\mathrm{Res}_1(\mathcal{O}_P,\mathcal{T})$ its first Fitting ideal is an Euler map of $\mathrm{Res}_1(\mathcal{O}_P,\mathcal{T})$ into the group of invertible fractional ideals of R. By 3.3, this map extends uniquely to an Euler map of $\mathrm{Res}(\mathcal{O}_P,\mathcal{T})$. Moreover, if $\mathcal{O}_F$ denotes the subset of $\mathcal{M}$ corresponding to the finitely generated free modules, then $\mathrm{Res}_1(\mathcal{O}_F,\mathcal{T})$ is mapped into the group of <u>principal</u> invertible fractional ideals, and hence $\mathrm{Res}(\mathcal{O}_F,\mathcal{T})$ is also mapped into this group.

<u>Remarks</u>. 1) The existence of the Euler map f of 3.4 is only a part of MacRae's paper [M]. His main theorem asserts that (when R is noetherian) the image under f actually consists of <u>integral</u> ideals of R. A similar assertion holds for the Euler map of the

first example of §1; see [KCR,p.140, Theorem 192]. David Rush and I, and independently, Y. Quentel, have shown that the noetherian hypothesis is unnecessary for MacRae's theorem.

2) Note that if one interprets the axioms $\mathscr{S}1$ - $\mathscr{S}7$ for exact sequences of isomorphism classes of R-modules by inserting arrows $\leftarrow$ instead of $\rightarrow$, then the axioms remain valid.

References

[B] H. Bass, Algebraic K-Theory, W.A. Benjamin, Inc., New York, 1968.

[KCR] I. Kaplansky, Commutative Rings, Allyn & Bacon, Boston, 1970.

[K] I. Kaplansky, Commutative Rings, Lecture Notes, Queen Mary College, London, 1966.

[M] R. MacRae, On an application of the Fitting invariants, J. of Algebra 2(1965), 153-169.

[Mi] B. Mitchell, Theory of Categories, Academic Press, New York, 1965

[SE] R. Swan and E.G. Evans, K-Theory of Finite Groups and Orders, Lecture Notes in Math. 149, Springer-Verlag, Berlin, 1970.

CHAIN CONJECTURES AND H-DOMAINS

Louis J. Ratliff, Jr.[1]
University of California
Riverside, California 92502

Some new equivalences to the chain conjecture (the integral closure of a local domain is catenary) and to the catenary chain conjecture (the integral closure of a catenary local domain is catenary) are proved, as are some new characterizations of a local H-domain. Also, a fact which lends support to the chain conjecture is noted, and it is proved that the H-conjecture (a local H-domain is catenary) implies the catenary chain conjecture.

1. INTRODUCTION AND TERMINOLOGY

To start with, a number of definitions are needed. These will be given only for an integral domain A , since our main interest in this paper is with integral domains.

(1.1) A chain of prime ideals $P_o \subset P_1 \subset \cdots \subset P_k$ in A is a maximal chain of prime ideals in A in case $P_o = (0)$, P_k is a maximal ideal in A , and, for each $i = 1,\ldots,k$, there is no prime ideal P in A such that $P_{i-1} \subset P \subset P_i$ (that is, height $P_i/P_{i-1} = 1$). The length of the chain is k .

(1.2) A satisfies the first chain condition for prime ideals (f.c.c.) in case each maximal chain of prime ideals in A has length equal to the altitude of A .

(1.3) A is catenary (or, satisfies the saturated chain condition for prime ideals) in case, for each pair of prime ideals $P \subset Q$ in A, $(A/P)_{Q/P}$ satisfies the f.c.c.

(1.4) A satisfies the second chain condition for prime ideals (s.c.c.) in case each integral domain B which is integral over A satisfies the f.c.c.

(1.5) A satisfies the chain condition for prime ideals (c.c.)

[1]Research on this paper was supported in part by the National Science Foundation, Grant 28939.

in case, for each pair of prime ideals $P \subset Q$ in A, $(A/P)_{Q/P}$ satisfies the s.c.c.

There are a number of relations between these last four definitions. Many of these relations can be found in [5, Remarks 2.5-2.7] and [7, Remarks 2.22-2.25], and these lists are sufficient for the needs of this paper.

With the above definitions, we can now state the chain conjecture.

(1.6) CHAIN CONJECTURE: The integral closure of a local (Noetherian) domain satisfies the c.c.

This conjecture is equivalent to: The integral closure of a local domain is catenary. Some further equivalent formulations of the conjecture were given by M. Nagata in 1956 in [3, Problems 3, 3', and 3", p. 62] (see (2.4) below).

The history of the chain conjecture can be traced back at least to 1937. It was then that W. Krull in [2, p. 755] asked if the following condition (which is (formally) weaker than the chain conjecture, when A is the integral closure of a Noetherian domain) holds: If $A \subset B$ are integral domains such that A is integrally closed and B is integral over A, and if $P \subset Q$ are prime ideals in B such that height $Q/P = 1$, then is it necessarily true that height $(Q \cap A)/(P \cap A) = 1$? I. Kaplansky in 1972 in [1] showed that the answer is "no." In the example given, however, A wasn't the integral closure of a Noetherian domain (nor a Krull domain). Just prior to this, J. Sally [9] gave an example of an integrally closed quasi-local domain which isn't catenary. Again, this example wasn't a Krull domain. Of course, each of these examples shows that the chain conjecture doesn't hold for arbitrary integrally closed domains.

Very little progress has been made on proving (or, disproving) the chain conjecture. In Section 2 of this paper, some new equivalences to this conjecture are given in (2.4) and, using one of these, some empirical evidence which supports the chain conjecture is given in (2.7) and (2.8) (see the comment prior to 2.8). In Section 3, some characterizations of a local H-domain (Definition 2.1) are given in (3.1) and (3.2). In Section 4, (3.2) is used to show that, of two other conjectures which have appeared in the literature (the H-conjecture (4.1) and the catenary chain conjecture (4.2)), the H-conjecture implies the catenary chain conjecture. A new formulation of the catenary chain conjecture is given in (4.7), and this paper is closed with some remarks on this formulation.

The undefined terminology in this paper is the same as that in [4].

2. THE CHAIN CONJECTURE AND H-DOMAINS

To give some new equivalences to the chain conjecture, the following definition is needed.

(2.1) DEFINITION. An integral domain A is an <u>H-domain</u> in case, for each height one prime ideal p in A, depth p = altitude A - 1.

The condition defining an H-domain originated in the study of catenary local domains. That is, the statement 'A local domain is an H-domain' is a kind of "dual" to statement (3) in the following known result [8, Theorem 2.2 and Remark 2.6(i)]:

(2.2) The following statements are equivalent for a local domain R :

(1) R is catenary.

(2) For each prime ideal P in R, height P + depth P = altitude R.

(3) For each depth one prime ideal P in R, height P = altitude R - 1.

One reason for the name "H-domain" is that the condition defining the ring has to do with <u>height</u> one prime ideals. Thus, perhaps a better name for such a ring would be an H_1-domain. Then an H_n-domain (for

$n \geqq 1$) could be defined analogously. However, the main reason for the name H-domain is it is hoped that every Henselian local domain is an H-domain (see (2.4) below).

(2.3) REMARK. For the remainder of this article, the following notation is fixed: R denotes a local domain, M is its maximal ideal, a = altitude R, F is the quotient field of R, and R' is the integral closure of R in F.

In the following theorem, the equivalence of (3), (4), (3'), and (4') was stated in [3, p. 62].

(2.4) THEOREM. The following statements are equivalent:

(1) The chain conjecture holds.

(2) Every local domain R such that R' is quasi-local is an H-domain.

(3) Every local domain R as in (2) is catenary.

(4) Every local domain R as in (2) satisfies the s.c.c.

(2') Every Henselian local domain is an H-domain.

(3') Every Henselian local domain is catenary.

(4') Every Henselian local domain satisfies the s.c.c.

Proof. (1) ⇒ (4), by [7, Remark 2.23(iv)], clearly (4) ⇒ (3) ⇒ (2), and (2) ⇒ (2'), by [4, (43.12)]. Since R/p is a Henselian local domain, if R is and p is a prime ideal in R, it follows by induction on $a \geqq 1$ that (2') ⇒ (3'). Also, it is known [6, Theorem 2.21] that (3') ⇔ (4') ⇔ (4) (since the Henselization of a local domain as in (2) is a Henselian local domain [4, (43.10), (43.11), and (43.20)]). Finally, (4) implies (1), since, if M' is a maximal ideal in R' and b is an element in M' which is not in any other maximal ideal in R', then $R'_{M'}$ is the integral closure of $R[b]_{M' \cap R[b]}$, hence satisfies the s.c.c., by (4), and so R' satisfies the c.c. [7, Remark 2.23(iv)], q.e.d.

Before proceeding, the following results on quasi-local H-domains are needed.

(2.5) REMARK. Let (S,N) be a quasi-local domain which is integral over a local domain R, and let altitude $S = a > 1$. Then the following statements are equivalent [7, Lemma 4.6, Proposition 4.7, and Corollary 4.10]:

(1) R is an H-domain.

(2) S is an H-domain.

(3) For all analytically independent elements b,c in S, height $NS[c/b] = a-1$. (It is known that $NS[c/b]$ is a depth one prime ideal in $S[c/b]$ [5, Lemma 4.3].)

Moreover, if S is integrally closed in its quotient field K, then each of these statements is equivalent to:

(4) For each $x \in K$ such that neither x nor $1/x$ is in S, height $NS[x] = a-1$.

By the equivalence of (1) and (2) in (2.4), a new approach in investigating the chain conjecture is available. Namely, attempt to prove:

(2.6) If, for some element x in F, $MR[x]$ is proper and height $MR[x] < a-1$, then R' isn't quasi-local.

If (2.6) can be proved, then whenever R' is quasi-local, necessarily R is an H-domain, by (1) and (3) for R in (2.5), hence the chain conjecture holds, by (1) and (2) in (2.4).

The following lemma is related (in spirit) to (2.5), as will be explained following its proof.

(2.7) LEMMA. _Assume_ R _isn't an H-domain, let_ p _be a height one prime ideal in_ R _such that_ $d = \text{depth } p < a-1$, _and let_ $0 \neq c \in p$. _Then there exists an element_ $b \in M$ _such that_ b,c _are analytically independent in_ R _as are_ b^2,c _and, with_ $x = b^2/c$ _and_ $y = 1/x = c/b^2$, height $MR[x] =$ height $MR[y] <$ height $(M,y)R[y] = d+1 <$ height $(M,x)R[x] = a$.

Proof. By the proof of [7, Proposition 4.7 (last paragraph)], there exists an element $b \in M$ such that: (i) $cR:bR = cR:b^2R$ is the p-primary component of cR ; and, (ii) b,c are analytically independent in R , as are b^2,c . Let $x = b^2/c$ and $y = c/b^2$. Then $P = (M,y)R[y]$ is a maximal ideal in R[y] and height P = d+1 (see the proof of [7, Proposition 4.7 (third paragraph)]). Also, $M^* = MR[x]$ and $M^{**} = MR[y]$ are depth one prime ideals [5, Lemma 4.3] and height M^* = height MR[x,y] = height M^{**}. Further, Q = (M,x)R[x] is a maximal ideal in R[x] and height Q = a , as will now be shown. Clearly $xR[x] \cap R \subseteq xR'[x] \cap R' \cap R = (K,X)R'[X] \cap R' \cap R$ (where K is the kernel of the natural homomorphism from R'[X] onto R'[x]) = (by [4, (11.13)]) $(b^2R':cR',X)R'[X] \cap R' \cap R = (b^2R':cR') \cap R = (b^2R' \cap R):cR$. Let $I = (b^2R' \cap R):cR$, and assume it is known that $I \subseteq \operatorname{Rad} bR$. Then, with A = R[x]/xR[x] , altitude A ≧ altitude R/bR = a-1 . Also, since $A = R/(xR[x] \cap R)$ is a local ring, Q is the only maximal ideal in R[x] which contains x . Therefore, altitude A = altitude $R[x]_Q/xR[x]_Q$ = height Q - 1 . Hence, since height Q ≦ a (since R satisfies the altitude inequality [5, Remark 2.11(i)]), it follows that height Q = a .

Thus it remains to show that $I \subseteq \operatorname{Rad} bR$. Let q be a height one prime ideal in R such that $b \in q$, so clearly $b^2R' \cap R \subseteq q$. If $c \notin q$, then $I \subseteq IR_q \cap R = (b^2R' \cap R)R_q \cap R \subseteq q$. If $c \in q$, then $q \neq p$, and so b is in the q-primary component of cR (by (i)). Therefore $x \in qR_q$, hence $1/x = c/b^2$ is not in the integral closure $(R_q)'$ of R_q . Thus $(b^2(R_q)':c(R_q)') \cap R_q \subseteq qR_q$, and this implies $I \subseteq ((b^2(R_q)':c(R_q)') \cap R \subseteq qR_q \cap R = q$. Therefore $I \subseteq \operatorname{Rad} bR$, as desired, q.e.d.

(2.7) shows that for every local domain R which isn't an H-domain, there exist elements $x \in F$ such that, with A = R[x]: (a) Height MA < a-1 ; and, (b) There exists a maximal ideal N in A such that $MA \subset N$ and height N > height MA + 1 . Also, it is known [7, Remark 4.4(iv)] that: (c) There exist at most a finite number of such N as in (b). (2.8) below shows that if R isn't an H-domain and R' satisfies the c.c. (or is catenary), then, for all such $x \in F$, (a) happens because there exists a maximal ideal M'' in R' such that height M'' = height $MA + 1$, and, for each N as in (b), there exists a maximal ideal M' in R' such that $x \in R'_{M'}$ and $M'R'_{M'} \cap A = N$. Of course, (c) is accounted for by the fact that there are always only finitely many maximal ideals in R'. Therefore, in summary, if the chain conjecture holds, then (a) - (c) happen because of conditions in R'. Of course, this in no way indicates how to prove (2.6), but (2.7) and (2.8) together lend empirical support to the chain conjecture.

(2.8) PROPOSITION. _Assume_ R _isn't an_ H-_domain, let_ $x \in F$ _such that, with_ $A = R[x]$, h = height $MA + 1 <$ height $(M,x)A$, _and let_ $\mathcal{P} = \{M' ; M'$ _is a maximal ideal in_ $R'\}$.

(1) _If there exists_ $M' \in \mathcal{P}$ _such that_ $R'_{M'}$ _is an_ H-_domain, then_ R' _isn't quasi-local_.

(2) _Assume there exists_ $M' \in \mathcal{P}$ _such that_ $R'_{M'}$ _is an_ H-_domain and_ height $M' > h$. _Then_ x _or_ $1/x$ _is in_ $R'_{M'}$ (_say_, x) _and_ $N = M'R'_{M'} \cap A$ _is a maximal ideal in_ A _such that_ $MA \subset N$ _and_ height $N > h$.

(3) _Assume every_ $M' \in \mathcal{P}$ _is such that_ $R'_{M'}$ _is an_ H-_domain_. _Then, for each maximal ideal_ N _in_ A _such that_ $MA \subset N$ _and_ height $N > h$, _there exists_ $M' \in \mathcal{P}$ _such that_ $x \in R'_{M'}$ _and_ $N = M'R'_{M'} \cap A$. _Moreover, there exists a maximal ideal_ M'' _in_ R' _such that_ height $M'' = h$.

Before proving (2.8), it should be noted that, since R isn't an H-domain, there exist such $x \in F$ (2.7). Further, since $R[x,1/x]$ is a quotient ring of $R[x]$ and of $R[1/x]$, it is clear that there exists a one-to-one correspondence between the ideals N in (3) such that $N \neq (M,x)A$ and the maximal ideals $N^* \neq (M,1/x)R[1/x]$ in $R[1/x]$ such that $MR[1/x] \subset N^*$ and height $N^* > h$.

Proof. (1) follows from (2.5). For (2), if neither x nor $1/x$

is in $R'_{M'}$, then $M'^* = M'R'[x]$ is a depth one prime ideal [10, Corollary, p. 20], and height $M'^* =$ height $M'R'_{M'}[x] \geqq h$, by hypothesis and (2.5). This, by integral dependence, implies the contradiction $h \leqq$ height $M'^* \leqq$ height MA ($M'^* \cap A = MA$, since $R'[x]/M'^*$ is integral over $A/(M'^* \cap A)$). Therefore x or $1/x$ is in $R'_{M'}$, say $x \in R'_{M'}$. Let $N = M'R'_{M'} \cap A$. Then, by integral dependence of $R'[x]$ over A, N is a maximal ideal in A, $N \cap R = M$ (so $MA \subset N$), and height $N \geqq$ height $M' > h$.

For (3), let N be a maximal ideal in A such that $MA \subset N$ and height $N > h$, let N' be a maximal ideal in $B = R'[x]$ such that $N' \cap A = N$ and height $N' =$ height N, and let $M' = N' \cap R'$, so M' is a maximal ideal in R'. Then x or $1/x$ is in $R'_{M'}$, for, if not, then $M'B$ is a depth one prime ideal in B which lies over MA, so height $M' - 1 =$ (2.5) height $M'B \leqq$ height MA, and so height $MA + 1 \geqq$ height $M' \geqq$ (by the altitude inequality) height N'. But height $N' =$ height $N >$ height $MA + 1$. Hence x or $1/x$ is in $R'_{M'}$. If $1/x \in M'R'_{M'}$, then $R'[1/x] \subseteq R'_{M'} \leqq B_{N'}$, and so $1/x \in N'B_{N'}$, which is a contradiction. Hence $x \in R'_{M'}$, and so $R' \subseteq B \subseteq R'_{M'}$ implies $M'R'_{M'} \cap B = N'$. Therefore $M'R'_{M'} \cap A = N$. Finally, depth $MA > 0$ implies MA is a depth one prime ideal, and so there is a maximal ideal M'' in R' such that, with $P = M''R'[x]$, depth $P = 1$, $P \cap A = MA$, and height $P =$ height MA. Hence, by (2.5), height $M'' =$ height $MA + 1 = h$, q.e.d.

3. SOME CHARACTERIZATIONS OF H-DOMAINS

Because of the equivalence of (1) and (2) in (2.4), it is of some interest to study H-domains. Some further reasons to study H-domains are indicated in (4.1) and (4.5) below. For these reasons, some characterizations of an H-domain will be given in this section.

<u>The fixed notation of</u> (2.3) <u>will continue to be used throughout this section</u>.

The main reason for including the following theorem is because it lends support to the H-conjecture (4.1), since it is known that if R is catenary, then R^* is catenary [7, Theorem 4.11].

(3.1) THEOREM. Let X be an indeterminate, and let $R^* = R[X]_{MR[X]}$. Then R is an H-domain if and only if R^* is an H-domain.

Proof. Assume first that R is an H-domain and let p^* be a height one prime ideal in R^*. If $p = p^* \cap R \neq (0)$, then height $p = 1$ and $p^* = pR^*$, hence depth p^* = altitude R^*/pR^* = altitude R/p = depth $p = a-1$, as desired. If $p^* \cap R = (0)$, then $a > 1$ and $x = X$ modulo $(p^* \cap R[X])$ is algebraic over R. Thus there is a non-zero element $r \in M$ such that rx is integral over R, hence $S = R[r^2x]$ is a local domain which is integral over R and whose quotient field is the quotient field of $R[x]$. Hence S is an H-domain (2.5). Let $N = (MR^*/p^*) \cap S$, so height $N = a$. Also, $NS[x] = MR[X]/(p^* \cap R[X])$ is a depth one prime ideal, so r^2, r^2x are analytically independent elements in S [6, Lemma 4.3], hence height $NS[x] = a-1$ (2.5). Thus depth $p^* = a-1$, and so R^* is an H-domain.

Conversely, let R^* be an H-domain and let p be a height one prime ideal in R. Then pR^* is a height one prime ideal in R^*, so depth p = (as in the first part of this proof) depth $pR^* = a-1$, hence R is an H-domain, q.e.d.

The following theorem will be used in Section 4.

(3.2) THEOREM. Let $P = R[X]_{(M,X)R[X]}$, where X is an indeterminate. Then the following statements are equivalent:

(1) R is an H-domain and $R^{(1)} \subseteq R'$, where $R^{(1)} = \cap\{R_p ; p$ is a height one prime ideal in $R\}$.

(2) R' is an H-domain.

(3) P is an H-domain.

Proof. It is known [6, Corollary 5.7(1)] that $R^{(1)} \subseteq R'$ if and only if, for each height one prime ideal p' in R', height $p' \cap R = 1$. Therefore, since depth p' = depth $p' \cap R$ (by integral dependence), (1) implies (2).

Assume (2) holds and let p^* be a height one prime ideal in P. If $p = p^* \cap R \neq (0)$, then height $p = 1$ and $p^* = pP$, hence depth p^* = depth $p+1 = a$ (since, by the Lying-Over Theorem, (2) implies R is an H-domain). If $p^* \cap R = (0)$, then let $L = R[x]_{(M,x)R[x]}$, where $x = X$ modulo $(p^* \cap R[X])$, so $L = P/p^*$ and it remains to show that altitude $L = a$. Since x is algebraic over R, there is a local domain $S = R[r^2x] \subseteq R[x] \subseteq L$ such that S is integral over R and the quotient fields of S and L are the same (as in the proof of (3.1)). Let S' be the integral closure of S. Since $x = c/b$ with b and c in S, $S'[x]/xS'[x] \cong S'/(cS':bS')$ (by [4, (11.13)], as in the proof of (2.7)). Now (2) implies S' is an H-domain (by integral dependence and [4, (10.14)]), hence depth $xS'[x]$ = depth $cS':bS'$ = altitude $S' - 1 = a-1$. Therefore, depth $xR[x] = a-1$, since $R[x] = S[x]$. Hence, since $(M,x)R[x]$ is the only maximal ideal in $R[x]$ containing x, altitude $L = a$. Hence (3) holds.

If (3) holds, then that R is an H-domain follows as in the last of the proof of (3.1). To prove that $R^{(1)} \subseteq R'$, it suffices to prove that if p' is a height one prime ideal in R', then height $p' \cap R = 1$ [6, Corollary 5.7(1)]. Suppose height $p' \cap R > 1$. Then it is known [7, Proposition 2.11] that either p' is a height one maximal ideal or there exists a height one prime ideal p in R such that depth p = depth p' = depth $p' \cap R < a-1$. Hence, since R is an H-domain, p' is a height one maximal ideal in R'. Therefore, with $x \in p'$ such that $1-x$ is in all the other maximal ideals in R', altitude $R[x]_{(M,x)R[x]} = 1 < a$. But this implies that P isn't an H-domain. Hence height $p' \cap R = 1$, and so (3) implies (1), q.e.d.

(3.3) COROLLARY. _Let_ R _be an_ H-_domain such that_ $R^{(1)} \subseteq R'$, *and let* $R_n = R[X_1,\ldots,X_n]$, *where* $X_1,\ldots,X_n$ *are algebraically independent over* R . *Then, for each prime ideal* P *in* R_n *such that* $P \cap R = M$, R_{nP} *is an H-domain*.

Proof. The proof is by induction on n . If $n = 1$, then by (3.1) it may be assumed that $MR_1 \subset P$, hence $P = (M,f)R_1$, for some monic polynomial $f \in P$. Then R_1 is integral over $A = R[f]$ (hence f is transcendental over R) and $Q = P \cap A = (M,f)A$, so A_Q is an H-domain (3.2). Hence, since $QR_1 = P$, it follows from (2.5) that R_{1P} is an H-domain. Using the case for $n = 1$, the remainder of the proof is straightforward, q.e.d.

(3.4) REMARK. (1) If R is an H-domain, then either $R^{(1)} \subseteq R'$ or there exists a height one maximal ideal in R' (as in the proof that (3) implies (1) in (3.2)).

(2) If R is an H-domain and $R^{(1)} \subseteq R'$, then every maximal ideal in R' has height a . For, if M' is a maximal ideal in R' such that height $M' < a$, then (by (1)) there exists a height one prime ideal $p' \subset M'$ such that depth $p' < a-1$. Since R is an H-domain, height $p' \cap R > 1$. But this and [7, Proposition 2.11] imply the contradiction R isn't an H-domain.

4. ON THE H-CONJECTURE AND THE CATENARY CHAIN CONJECTURE

In this section, (3.2) will be used to show an implication between two conjectures which have appeared in the literature. Then, a new formulation of one of these conjectures will be given.

The notation of (2.3) will continue to be used throughout this section.

(4.1) H-CONJECTURE: If R is an H-domain, then R is catenary.

The converse of the conjecture clearly holds.

(4.2) CATENARY CHAIN CONJECTURE: If R is catenary, then R' satisfies the c.c.

This conjecture is equivalent to: If R is catenary, then R'

is catenary. Some other equivalences are given in the following theorem.

(4.3) THEOREM. If R is catenary, then the following statements are equivalent:

(1) R' satisfies the c.c.

(2) Every integral domain which is a finite R-algebra is catenary.

(3) For each height one prime ideal p in R, R/p satisfies the s.c.c.

Proof. If $a = 1$, then the theorem is easy, so it will be assumed that $a > 1$.

Assume (1) holds, let A be an integral domain which is a finite R-algebra, and let N be a maximal ideal in A. To prove that A is catenary it suffices to prove that $B = A_N$ is catenary. For this, it is known that height $N = 1$ or a [7, Proposition 3.1(1)], so clearly it may be assumed that height $N = a > 1$. Let P be a depth one prime ideal in B, let $C = R'[A]$, and let $D = C_{(A \sim N)}$. Then D is integral over B and D is catenary, by (1). Let P' be a prime ideal in D such that $P' \cap B = P$ and height P' = height P. Then depth $P' = 1$ and, if Q is a maximal ideal in D such that $P' \subset Q$, then $Q \cap R = M$, hence height Q = height $Q \cap C = a$ [7, Proposition 3.1(1)]. Hence height P = height $P' = a-1$, and so B is catenary (2.2).

Assume (2) holds, let p be a height one prime ideal in R, let $\overline{R} = R/p$, and let $\overline{A}$ be an integral domain which is a finite $\overline{R}$-algebra. To prove that $\overline{R}$ satisfies the s.c.c. it suffices to prove $\overline{A}$ satisfies the f.c.c. [5, Theorem 3.11]. For this, there is an integral domain A which is a finite R-algebra such that $A/p^* = \overline{A}$, for some height one prime ideal p^* in A. Then (2) implies $\overline{A}$ is catenary, and it follows from [7, Proposition 3.1(1)] (since R, A, and $\overline{R}$ are catenary) that every maximal ideal in $\overline{A}$ has height a-1.

Hence (3) holds.

Assume (3) holds, and assume, at first, that R' has no height one maximal ideals. Then it is known that R satisfies the s.c.c. [5, Proposition 3.3], and so R' satisfies the s.c.c. Thus it remains to consider the case when R' has a height one maximal ideal. In this case, Lemma 4.4 below shows that, for some $x \in R'$, $L = R[x,1/x]$ is a catenary local domain whose integral closure $R'[1/x]$ has no height one maximal ideals and the only prime ideals in R' which blow up in $R'[1/x]$ are the height one maximal ideals. Thus, clearly R' satisfies the c.c., if $R'[1/x]$ satisfies the s.c.c.; and, as already noted, $R'[1/x]$ satisfies the s.c.c., if (3) holds for L. So, let p" be a height one prime ideal in L, let $p' = p'' \cap R[x]$, and let $p = p' \cap R$. Then height p = 1 [7, Proposition 3.1(1)], so R/p satisfies the s.c.c. (by (3)), hence, by integral dependence, $R[x]/p'$ satisfies the s.c.c. Finally, $L/p'' = R[x]/p'$, since, by the proof of (4.4), R[x] has only two maximal ideals $N_1 = (M,x)$ and $N_2 = (M,1-x)$ and $p' \subset N_2, \not\subseteq N_1$, q.e.d.

Another equivalence of the catenary chain conjecture will be given in (4.7) below.

(4.4) LEMMA. _Assume_ R _is catenary and_ $a > 1$, _and let_ A _be an integral domain which is integral over_ R. _If there is a maximal ideal_ N _in_ A _such that_ height $N < a$, _then there exists an element_ $x \in R'$ _such that_:

(1) $R[x,1/x]$ _is a catenary local domain_;

(2) _Every maximal ideal in_ $A[x,1/x]$ _has_ height a ; _and_,

(3) _The prime ideals in_ R' _which blow up in_ $R'[1/x]$ _are the height one maximal ideals_.

Proof. If such N exist, then necessarily height N = 1 [7, Proposition 3.1(1)]. Let $B = R'[A]$, let Q' be a maximal ideal in B such that $Q' \cap A = N$, and let $Q = Q' \cap R'$. Then Q is a maximal ideal in R' and height Q = 1 [4, (10.14)]. So, let x be an element which is in a prime ideal in R' if and only if the ideal is a height one maximal ideal, and let y be an element in all other maximal ideals in R' such that $x+y = u$ is a unit in R'. Clearly it may be assumed that $u = 1$. Then $R_1 = R[x]$ has exactly two maximal ideals, namely, $N_1 = (M,x)R_1$ and $N_2 = (M,1-x)R_1$, and height $N_1 = 1$ and height $N_2 = a$, so $L = R[x,1/x] = R_{1N_2}$ is a local domain. Also, $R'[1/x]$ is the integral closure of L (so (3) holds), and $B[1/x]$ is integral over $R'[1/x]$ and over $A[x,1/x]$. Thus, it follows from the choice of x and [4, (10.14)] that $A[x,1/x]$ has no height one maximal ideals. Therefore, if (1) holds, then (2) follows from [7, Proposition 3.1(1)].

To prove that L is catenary, let $C = R + (N_1 \cap N_2)$, so C is a local domain with maximal ideal $J = N_1 \cap N_2$, and $R \subseteq C \subset R_1 = R[x]$. Let P' be a depth one prime ideal in L, let $P = P' \cap R_1$, and let $p = P \cap C$. Then $L_{P'} = R_{1P} = C_p$ (since J is the conductor of the integral closure of C in R_1), hence height P' = height P = height p. Further, depth P' = depth P (= height N_2/P), and depth P = depth p. Hence, since C is catenary [7, Theorem 3.2], height P' = height $p = a-1$, and so L is catenary (2.2), q.e.d.

(4.5) THEOREM. _If the H-conjecture holds, then the catenary chain conjecture holds._

Proof. Let R be a catenary local domain. To prove that R' satisfies the c.c. it suffices to prove that, for each maximal ideal M' in R', $R'_{M'}$ satisfies the s.c.c. [7, Remark 2.23(iv)]. Clearly this holds if height $M' = 1$, so by (4.4) it may be assumed that every M' has height $a > 1$. Let p' be a height one prime ideal in R'. If height $p' \cap R > 1$, then depth $p' < a-1$, hence, by [7, Proposition 2.11], R isn't an H-domain. This is a contradiction, so height $p' \cap R = 1$, hence $R^{(1)} \subseteq R'$ [6, Corollary 5.7(1)] and R is an H-domain. Therefore $P = R[X]_{(M,X)R[X]}$ is an H-domain, by (3.2), hence, by hypothesis, P is catenary. But it is known

[6, Theorem 2.21] that P is catenary (if and) only if R satisfies the s.c.c. Hence R' satisfies the s.c.c., q.e.d.

(4.6) COROLLARY. If the H-conjecture holds, then every integrally closed catenary local domain satisfies the s.c.c.

Proof. This follows from (4.5), since a local domain satisfies the s.c.c. if and only if it satisfies the c.c. [7, Remark 2.23(v)], q.e.d.

(4.7) indicates another approach to proving the catenary chain conjecture which might prove more accessible.

(4.7) THEOREM. The catenary chain conjecture holds if and only if the following condition holds: If R is catenary and $A = R[x]$ is such that A has exactly two maximal ideals $N_1 = (M,x)$ and $N_2 = (M,1-x)$ and $N_1 \cap N_2 = M$ is the maximal ideal in R, then A is catenary.

Proof. By (4.3) the catenary chain conjecture implies the condition, so assume that R is catenary and the condition holds. If altitude $R = a = 1$, then R satisfies the s.c.c., hence it may be assumed that $a > 1$. By (4.3), it suffices to prove that if B is an integral domain which is a finite integral extension of R, then B is catenary.

For this, if B is local, then B is catenary [7, Theorem 3.2], so it may be assumed that B isn't local. Clearly, if N is a height one maximal ideal in B, then B_N is catenary. Hence, by (4.4) it may be assumed that every maximal ideal in B has height a.

Let N be a maximal ideal in B, and let $x \in N$ such that $1-x$ is in all the other maximal ideals in B. Let $A = R[x]$. Then $N_1 = (M,x)A$ and $N_2 = (M,1-x)A$ are the maximal ideals in A. Let $R_1 = R + (N_1 \cap N_2)$, so R_1 is a catenary local domain [7, Theorem 3.2]. Therefore, by hypothesis, A is catenary, so to prove that B is catenary, it suffices to prove that $B_{(A\sim N_2)}$ is catenary (since $B_{(A\sim N_1)} = B_N$ is integral over A_{N_1}, hence is catenary [7, Theorem

3.2]). Since A_{N_2} and $B_{(A\sim N_2)}$ satisfy the conditions on R and B, it follows by finitely many repetitions that B is catenary, q.e.d.

Call the type of extension ring in (4.7) special. An investigation of special extensions could prove rewarding, and, in particular, might give the needed results to show that A is catenary whenever R is. In the following remark, a proof of some of the more evident properties of special extensions will be given.

(4.8) REMARK. Let (R,M) be a (not necessarily catenary) local domain, and let $A = R[x]$ be a special extension of R. Then the following statements hold:

(1) For each element $y \in A, \notin R, A = R[y]$.

(2) If I is an ideal in A, then either $A/I = R/(I\cap R)$ (if $I \not\subseteq N_1 \cap N_2 = M$), or A/I is a special extension of $R/(I\cap R)$ (but not necessarily an integral domain).

(3) There is a one-to-one correspondence between the prime ideals $p \neq M$ in R and the prime ideals $P \notin \{N_1, N_2\}$ in A given by $R_p = A_P$. Moreover, for all prime ideals Q in A, $QA_Q = qA_Q$, where $q = Q\cap R$.

(4) It follows from (2) and (3) that A is locally unramified over R (see [4, pp. 144-145]).

(5) If $p \neq M$ is a prime ideal in R, then pA is semi-prime. In fact, if q is a p-primary ideal in R, then, with $q^* = qR_p \cap A$, either: (a) $qA = q^*$ (if $q^* \subset N_1 \cap N_2$); (b) $qA = q^* \cap N_1$ (if $q^* \subset N_2, \not\subseteq N_1$); or, (c) $qA = q^* \cap N_2$ (if $q^* \subset N_1, \not\subseteq N_2$). It follows in case (b) that $qA_{N_1} = N_1 A_{N_1}$, hence q^i isn't p-primary, for $i \geq 2$. A similar comment holds in case (c).

Proof. (1) follows from $A = R + xR$ and $M = R:A$, (2) follows from (1) and an easy computation, and (3) follows from $M = R:A$. For (5), if $q^* \subseteq N_1$, then $q^* \cap N_2 = q^* \cap M = q^* \cap R = q \subseteq qA \subseteq q^* \cap R$.

It follows that either $q = qA = q^{*} \cap N_2$ (if $q^{*} \subseteq N_1, \not\subseteq N_2$) or $q = qA = q^{*}$ (if $q^{*} \subseteq N_1 \cap N_2$), q.e.d.

BIBLIOGRAPHY

[1] I. Kaplansky, Adjacent prime ideals, J. Algebra 20(1972), 94-97.

[2] W. Krull, Beiträge zur Arithmetik kommutativer Integritätsbereiche. III Zum Dimensionsbegriff der Idealtheorie, Math. Zeit. 42(1937), 745-766.

[3] M. Nagata, On the chain problem of prime ideals, Nagoya Math. J. 10(1956), 51-64.

[4] M. Nagata, Local Rings, Interscience Tracts 13, Interscience, New York, 1962.

[5] L. J. Ratliff, Jr., On quasi-unmixed local domains, the altitude formula, and the chain condition for prime ideals, (I), Amer. J. Math. 91(1969), 508-528.

[6] L. J. Ratliff, Jr., On quasi-unmixed local domains, the altitude formula, and the chain condition for prime ideals, (II), Amer. J. Math. 92(1970), 99-144.

[7] L. J. Ratliff, Jr., Characterizations of catenary rings, Amer. J. Math. 93(1971), 1070-1108.

[8] L. J. Ratliff, Jr., Catenary rings and the altitude formula, Amer. J. Math., (forthcoming).

[9] J. Sally, Failure of the saturated chain condition in an integrally closed domain, Abstract 70T-A72, Notices Amer. Math. Soc. 17(1970), 560.

[10] O. Zariski and P. Samuel, Commutative Algebra, Vol. II, Van Nostrand, New York, 1960

ON THE NUMBER OF GENERATORS OF IDEALS OF DIMENSION ZERO

Judith Sally

Rutgers University

ABSTRACT. This note contains generalizations of two results of Abhyankar [1] which give a bound for the embedding dimension of certain local rings in terms of the multiplicity and the dimension of the ring.

Let R be a Noetherian ring. An ideal I has dimension 0 if the (Krull) dimension of R/I is 0. The multiplicity of an ideal I of dimension 0 is denoted $\mu(I)$; the length of an R-module A is denoted $\lambda(A)$. If (R, M) is a local ring and I is any ideal, v(I) denotes the number of elements in a minimal basis of I, i.e., v(I) is the dimension of I/IM as a vector space over R/M. When I = M, $\mu(M) = \mu(R)$ and v(M) is the embedding dimension of R.

Theorem 1. Let (R, M) be a d-dimensional local Macaulay ring. Let I be an M-primary ideal. Then,

$$v(I) \leq d + \mu(I) - \lambda(R/I).$$

Proof. We may assume $I \neq (0)$. The proof is by induction on d. If d = 0, then $\mu(I) = \lambda(R)$. Since $\lambda(R) \geq \lambda(R/I) + \lambda(I/IM)$, we have that $v(I) \leq \mu(I) - \lambda(R/I)$. Assume $d > 0$. By passing to R(X), MR(X) and IR(X) as in Nagata [3, p.18, p.71], we may assume that R/M is an infinite field. By Theorem 22.3 in Nagata [3], there is a non-zero divisor x in I but not in IM such that x is superficial for I. In the Macaulay ring R/xR we have, by induction, $v(I/xR) \leq d - 1 + \mu(I/xR) - \lambda((R/xR)/(I/xR))$. Since x is superficial and a non-zero divisor, $\mu(I/xR) = \mu(I)$. Since x is not in IM $v(I) = v(I/xR) + 1$. Thus, $v(I) \leq d + \mu(I) - \lambda(R/I)$.

Remark. Abhyankar's proof in [1] for the case I = M can be modified to give a proof of Theorem 1. However, since his proof uses a fact about systems of parameters in a local Macaulay ring which can be derived from Theorem 1, a different proof is given here.

The bound obtained in Theorem 1 is best possible. The following are examples of local Macaulay rings (R, M) with $v(M) = d + \mu(M) - 1$, i.e., $\mathrm{emdim}(R) = d + \mu(R) - 1$, and of an M-primary ideal I of R with $v(I) = d + \mu(I) - \lambda(R/I)$. Let K be a field. Let μ be any integer > 1.

For $d = 0$, take $R_\mu^0 = K[[x_1, \dots, x_{\mu-1}]]/(x_1, \dots, x_{\mu-1})^2$, where $x_1, \dots, x_{\mu-1}$ are indeterminates, and take I to be any ideal.

For $d = 1$, take $R_\mu^1 = K[[x^\mu, x^{\mu+1}, \dots, x^{2\mu-1}]]$, where x is an indeterminate, and take $I = (x^\mu, x^{\mu+1}, \dots, x^{2\mu-2})$.

For $d > 1$, take $R_\mu^d = R_\mu^1[[t_1, \dots, t_{d-1}]]$, where $t_1, \dots, t_{d-1}$ are indeterminates, and take $I = (x^\mu, x^{\mu+1}, \dots, x^{2\mu-2}, t_1, \dots, t_{d-1})$.

In order to give a global version of Theorem 1 we need the following definition. An ideal I of a ring R is <u>rank unmixed</u> if all primes minimal over I have the same rank (i.e., height).

<u>Corollary</u>. Let R be a d-dimensional Macaulay ring. Let I be a non-zero ideal of dimension 0 and rank r. If I is rank unmixed, then I can be generated by $\max(d + 1, r + \mu(I) - 1)$ elements. If R is semi-local, $\max(1, r + \mu(I) - 1)$ elements are enough.

<u>Proof</u>. We will show that I can be locally generated by $r + \mu(I) - 1$ elements. The desired conclusion will then follow from the Forster - Swan Theorem [4]. Let $P_1, \dots, P_k$ be the primes minimal over I. Since I has dimension 0, $IR_P = R_P$ for all primes $P \neq P_1, \dots, P_k$. By Theorem 1, $v(IR_{P_i}) \leq r + \mu(IR_{P_i}) - 1$, for $1, \dots, k$. Since I is rank unmixed, $\mu(I) = \sum_{i=1}^{k} \mu(IR_{P_i})$. Thus we have that $v(IR_{P_i}) \leq r + \mu(I) - 1$, for $i = 1, \dots, k$.

Abhyankar also proves in [1] that if (R, M) is a d-dimensional local ring such that R/M is algebraically closed and A, the associated graded ring of R, is a domain, then $\mathrm{emdim}(R) \leq d + \mu(R) - 1$. Theorem 2 below extends this result to M-primary ideals

<u>Theorem 2</u>. Let (R, M) be a d-dimensional local ring such that R/M is algebraically closed. Let I be an M-primary ideal. If the graded ring $G = R/M + I/IM + I^2/I^2M + \ldots$ is a domain, then

$$v(I) \leq d + \mu(I) - 1.$$

<u>Proof</u>. Assume $I \neq (0)$. Let H(G, n) denote the polynomial which, for large n, gives $\lambda(I^n/I^nM)$. H(G, n) is of degree t - 1, where t = dim G (cf. Zariski and Samuel [5, p.235]). Let $a/(t-1)!$ be the leading coefficient of H(G, n). Since R/M is algebraically closed, we have as in Abhyankar [1], that

$$a + t - 1 \geq \dim[I/IM : R/M] = v(I),$$

(also cf. Abhyankar [2, (12.3.5)]). To prove the theorem it is enough to show that t = d and that $\mu(I) \geq a$. Let $B = R/I + I/I^2 + \ldots$. B has dimension d. Now G = B/(M/I)B, so that (M/I)B is a prime homogeneous ideal of B. But, since $M^j \subseteq I$ for some positive integer j, we have that (M/I)B is the nilradical of B. Thus, t = dim G = dim B/(M/I)B = dim B = d. Let H(B, n) denote the polynomial which, for large n, gives $\lambda(I^n/I^{n+1})$. H(B, n) has degree d - 1 and leading coefficient $\mu(I)/(d-1)!$ Now $H(B, n) \geq H(G, n)$ because $\lambda(I^n/I^{n+1}) = \lambda(I^n/MI^n) + \lambda(MI^n/I^{n+1})$ and, since both polynomials have the same degree, we have that $\mu(I) \geq a$.

It is interesting to note that although the bound obtained in Theorem 2 is, in general, larger than the one in Theorem 1, it is best possible. Let (R, M) be the local, non-Macaulay domain of Nagata's Example 2 [3, p.203-205] for the case m = 0, r = 1. (We may take K algebraically closed.) Let xR' and N be the maximal ideals of

R', the integral closure of R. Then $M = xR' \cap N = xN$. Let $I = x^2N$. R, M and I satisfy the hypotheses of Theorem 2. $v(M) = v(I) = 2$; $d = 2$; $\mu(M) = \mu(I) = 1$. Hence, $v(M) = d + \mu(M) - 1$ and $v(I) = d + \mu(I) - 1$.

REFERENCES

1 Abhyankar, S. S., "Local rings of high embedding dimension", Amer. J. Math 89 (1967), 1073-1077.

2 Abhyankar, S. S., Resolutions of Singularities of Embedded Algebraic Surfaces, Academic Press, New York, 1966.

3 Nagata, M., Local Rings, Interscience, New York, 1962.

4 Swan, R. G., "The number of generators of a module", Math. Zeit. 102 (1967), 318-322.

5 Zariski, O. and Samuel, P., Commutative Algebra, vol. II, D. Van Nostrand, New York, 1960

RINGS OF GLOBAL DIMENSION TWO

by Wolmer V. Vasconcelos
Department of Mathematics, Rutgers University
New Brunswick, New Jersey 08903

This paper gives a change of rings theorem for homological dimensions which seems especially suited for the study of the commutative rings of global dimension two. Elsewhere we gave a description of the local rings of global dimension two that when applied to the global case yields an idea of what the spectrum of such rings look like "down" from a maximal ideal. Here an attempt is made to describe the spectrum "up" from a minimal prime ideal.

1. CHANGE OF RINGS IN DIMENSION TWO

As a general proviso, all rings considered here will be commutative. Bourbaki [3] will be the basic source for terminology and elementary notions.

I.Kaplansky in [12] proved the following result which is a simple but effective tool in the study of homological dimensions of rings and modules.

THEOREM 1.1. (*Change of rings in dimension one*) *Let* A *be any ring and* x *a central element of* A *which is neither a unit nor a nonzero divisor. Let* $B = A/(x)$ *and let* E *be a nonzero module over* B. *If the projective dimension* (*proj dim for short*) *of* E *over* B *is finite, then*

$$\text{proj dim}_A(E) = 1 + \text{proj dim}_B(E).$$

With A commutative this result was generalized by Cohen and Jensen [7,10] to the case of $B = A/I$ where I is a faithfully projec-

---

Partially supported by National Science Foundation Grant GP-19995.

tive ideal of A. Observe that such ideals are according to Vasconcelos [14] finitely generated. On the other hand, Gruson [8] proved recently that projectivity (and thus projective dimension) is a local property for the Zariski topology of Spec(A). In particular this remark implies that, in dimension one, it is enough to consider the change of rings with $I = (x)$.

In general, if E is a module over the ring $B = A/I$ one always has

(*) $$\text{proj dim}_A(E) \leq \text{proj dim}_B(E) + \text{proj dim}_A(B).$$

This follows for instance from the spectral sequence

(**) $$E_2^{p,q} = \text{Ext}_B^q(E, \text{Ext}_A^p(B,C)) \underset{p}{\Longrightarrow} \text{Ext}_A^n(E,C)$$

where C may be any A-module (cf. [5, XVI.5]).

Now we will consider the dimension two analogue of (1.1). Let I be a finitely generated ideal of A of projective dimension one. It is not always the case that equality in (*) holds. Thus, for instance, if A is a regular local ring of dimension $n > 1$ and $I = x(x,y)$ where x,y form a regular A-sequence, then equality holds in (*) if E is a finitely generated B-module (cf. [1, Cor 2.12]) but not for any module. In fact, as the Krull dimension of B is $n-1$, its finitistic projective dimension is also $n-1$ according to Gruson (cf. [8. p.82]) and thus we cannot have always equality in (*). Thus we must avoid this situation and also prevent, for more elementary reasons, I from sitting in a proper direct summand of A. Both conditions are skirted by asking that the canonical map $A \longrightarrow \text{Hom}_A(I,A)$ be an isomorphism. In fact, such ideal is faithful and if A is noetherian this implies that its grade is at least two. For brevity we call such ideals <u>overdense</u>.

THEOREM 1.2. (<u>Change of rings in dimension two</u>) <u>Let</u> A <u>be a commutative ring, let</u> I <u>be an overdense finitely generated ideal of projective dimension one and let</u> $B = A/I$. <u>If</u> E <u>is a nonzero</u> B-<u>module with finite</u>

<u>projective</u> <u>dimension</u> <u>over</u> B, <u>then</u>

$$\underline{\text{proj}}\ \underline{\text{dim}}_A(E) = 2 + \underline{\text{proj}}\ \underline{\text{dim}}_B(E).$$

Examples of such idealstend to abound in domains of global dimension two as we shall see. Other examples can be found in the following manner : Let A be a domain and let ϕ be an $n \times (n+1)$ matrix with entries in A. Let $I = D(\phi)$ be the ideal generated by the $n \times n$-minors of ϕ. If $I^{-1} = \{x \in K = \text{field of quotients of } A,\ xI \subset A\} = A$, then proj $\dim_A(I) = 1$ and I is overdense from Burch's theorem [4]. This is particularly easy to check in G.C.D. domains, e.g. with $A = k[x,y,z]$ let

$$\phi = \begin{bmatrix} x & y & z \\ y & z^2 & zy+1 \end{bmatrix}.$$

PROOF. We begin with some remarks.

(1) I is of finite presentation:

Let $$0 \longrightarrow K \longrightarrow A^n \longrightarrow I \longrightarrow 0$$

be a projective resolution of I. Thus K is projective and we claim it is also finitely generated. For that it is enough to show that for each prime ideal P the localized module K_P is free of rank $n-1$. Actually we have only to consider this for the minimal primes; let P be one such and localize the above sequence at P. As I is faithful, it cannot be contained in P and thus $I_P = A_P$ and thus $K_P = A_P^{n-1}$.

(2) K may be taken to be isomorphic to A^{n-1}:

First we pose that if $E = (0)$ then proj dim $(E) = -\infty$. By a theorem of Bourbaki (cf. [3, Chap.II, p.138]) there is a "partition of the unity" for which K is free, i.e. there exists $(f_1,\dots,f_m) = A$ with $K_f = A_f^{n-1}$, where K_f denotes localization of K at the multiplicative system of powers of $f = f_i$. While this is happening, for any $B = A/I$-module $E \neq (0)$ of finite projective dimension, will have that for some $f_i = f$ $E_f \neq (0)$ and proj $\dim_{B_f}(E_f)$ is the same as proj $\dim_B(E)$, according to the local characterization of projectivity. At the same time proj dim $(I_f) = 1$ since I_f A_f-projective would make it equal to A_f.

Assume then that we have an exact sequence

(***) $$0 \longrightarrow A^{n-1} \xrightarrow{\ \phi\ } A^n \longrightarrow I \longrightarrow 0.$$

(3) I is generated by the (n-1) x (n-1)-minors of ϕ:

Let $D = (d_1,\ldots,d_n)$ be the ideal generated by the minors of ϕ. Using the notation in Kaplansky's [11,p.148,Ex.8], we have $d_i a_j = d_j a_i$. Define the following map $\psi : I \longrightarrow D$ $\psi(\Sigma r_i a_i) = \Sigma r_i d_i$. First note that is well-defined as $\Sigma r_i a_i = 0$ yields $d_j(\Sigma r_i a_i) = (\Sigma r_i d_i) a_j = 0$ and thus $\Sigma r_i d_i = 0$ as I is faithful. On the other hand, since I is overdense ψ is realized by multiplication by an element d in A and $D = dI$. Since I is torsion-free it follows by McCoy's theorem that d is a unit in A and hence $I = D$.

(4) If C is a nonzero B-module $\mathrm{Tor}_2^A(B,C) \neq 0$:

Tensor (***) with $C \neq 0$ to get

$$0 \longrightarrow \mathrm{Tor}_1^A(I,C) \longrightarrow C^{n-1} \xrightarrow{\ \phi_C\ } C^n \longrightarrow I \otimes C \longrightarrow 0.$$

By (3) above $D = I$ implies that each (n-1)-minor of ϕ_C is annihilated by D and thus by McCoy's theorem $\mathrm{Tor}_1^A(B,C) = \ker\phi_C \neq 0$.

(5) For any nonzero B-module C $\mathrm{Ext}_A^2(B,C) \neq 0$:

Apply $\mathrm{Hom}_A(\ ,C)$ to (***) to get

$$0 \longrightarrow \mathrm{Hom}_A(I,C) \longrightarrow C^n \xrightarrow{\ {}^t\phi_C\ } C^{n-1} \longrightarrow \mathrm{Ext}_A^1(I,C) \longrightarrow 0$$

from which it readily follows that $\mathrm{Ext}_A^1(I,C) = \mathrm{Ext}_A^1(I,A) \otimes C$.

Now we claim that $T = \mathrm{Ext}_A^1(I,A)$ is a finitely generated faithful A/I-module. But this follows from the exact sequence

$$0 \longrightarrow \mathrm{Hom}_A(I,A) \longrightarrow A^n \longrightarrow A^{n-1} \longrightarrow \mathrm{Ext}_A^1(I,A) \longrightarrow 0$$

and the well-known companion exact sequence

$$0 \longrightarrow \mathrm{Ext}_A^1(\mathrm{Ext}_A^1(I,A),A) \longrightarrow I \longrightarrow \mathrm{Hom}_A(\mathrm{Hom}_A(I,A),A) \longrightarrow \mathrm{Ext}_A^2(\mathrm{Ext}_A^1(I,A),A).$$

The conclusion now follows from Gruson's result [8,p.58] that for such a module T, $T \otimes C = 0$ implies $C = 0$.

We are now ready to prove the theorem. Let E be a nonzero B-module with proj dim $_B(E) = n$. Notice that by (4) proj dim $_A(E) \geq 2$. The first case to examine is then

Proj dim $_B(E) = 1$:

Let
$$0 \longrightarrow F \longrightarrow F' \longrightarrow E \longrightarrow 0$$
be a projective resolution of E as a B-module. Set $C = \mathrm{Ext}^1_B(E,F)$ and remark $C \neq 0$. Consider the sequences
$$0 \longrightarrow \mathrm{Hom}_A(I,F) \longrightarrow F^n \longrightarrow L \longrightarrow 0$$
$$0 \longrightarrow L \longrightarrow F^{n-1} \longrightarrow \mathrm{Ext}^1_A(I,F) \longrightarrow 0$$
obtained by applying $\mathrm{Hom}_A(\ ,F)$ to (***) and then splicing. Applying $\mathrm{Hom}_B(E,\ )$ to both sequences and observing proj dim $_B(E) = 1$, we get the diagram
$$\begin{array}{ccccccc}
\mathrm{Ext}^1_B(E,\mathrm{Hom}_A(I,F)) & \longrightarrow & \mathrm{Ext}^1_B(E,F^n) & \longrightarrow & \mathrm{Ext}^1_B(E,L) & \longrightarrow & 0 \\
 & & & & \downarrow & & \\
 & & & & \mathrm{Ext}^1_B(E,F^{n-1}) & & \\
 & & & & \downarrow & & \\
 & & & & \mathrm{Ext}^1_B(E,\mathrm{Ext}^1_A(I,F)) & & \\
 & & & & \downarrow & & \\
 & & & & 0 & &
\end{array}$$
with exact row and column. Notice that $\mathrm{Ext}^1_B(E,\mathrm{Ext}^1_A(I,F)) = 0$ would mean that
$$[\mathrm{Ext}^1_B(E,F)]^n \xrightarrow{{}^t\phi\,\mathrm{Ext}(E,F)} [\mathrm{Ext}^1_B(E,F)]^{n-1} \longrightarrow 0$$
is exact. We know however that the cokernel of this map is $\mathrm{Ext}^1_A(I,A) \otimes \mathrm{Ext}^1_B(E,F)$ which by (5) is distinct from 0.

Passing now to the spectral sequence
$$E_2^{p,q} = \mathrm{Ext}^q_B(E,\mathrm{Ext}^p_A(B,F)) \underset{p}{\Longrightarrow} \mathrm{Ext}^n_A(E,F)$$
we see that $E_2^{p,q} = 0$ for $p > 2$, $q > 1$ but $E_2^{2,1} \neq 0$. Thus $\mathrm{Ext}^3_A(E,F) \neq 0$ and proj dim $_A(E) = 3$.

Proj dim $_B(E) = n > 1$:

Write
$$0 \longrightarrow L \longrightarrow F \longrightarrow E \longrightarrow 0$$
exact with F B-projective. Then proj dim $_A(F) = 2$ and by induction proj dim $_A(L) = n+1$. Thus, by dimension shifting, proj dim $_A(E) =$

n+2, as desired.

REMARK. It is clear how the procedure followed here yields the flat and injective analogues of the change of rings in dimension two.

2. FINITELY GENERATED IDEALS

Let A be, unless otherwise specified, a commutative ring of global dimension two. The great divider in the study of these rings seems to be the presence or lack of coherence, as in the former case the results of the preceding section provide a clear picture at least of the prime ideals sitting above overdense ideals. Examples where the coherence is missing have first been discovered by Jensen and Jøndrup; normally we will steer away from this case and limit ourselves to quoting [16, Prop. 3.4] :

PROPOSITION 2.1. A _is coherent if and only if_ Min(A), _the subspace of minimal prime ideals, is compact (This is also equivalent to "the total ring of quotients of_ A _is absolutely flat"_).

Portions of the structure of local rings of global dimension two is duplicated in coherent rings of finite global dimension with projective = free.

PROPOSITION 2.2. _Let_ A _be a commutative coherent ring such that every finitely generated ideal has finite projective dimension. If projectives are free (in particular if_ A _is local) then_ A _is a_ G.C.D. _domain_.

PROOF (Sketch). From Vasconcelos [15] it follows that such a ring is a domain. Let us now apply the method of MaRae [13] of associating to each finitely generated ideal I of finite projective dimension its "common divisor ideal", i.e. an ideal (d) such that I = dJ and J overdense. With I = (a,b), J = (e,f) and a = de, b = df we claim that proj dim (e,f) $\leq$ 1. Indeed, consider the sequence

$$0 \longrightarrow K \longrightarrow A^2 \xrightarrow{\phi} (e,f) \longrightarrow 0$$

with $\phi(x,y) = xe-yf$. If $(x,y) \in K$, i.e. $xe-yf = 0$, $x/f \in (e,f)^{-1} = A$ and hence $x \in Af$. Thus K is generated by (f,e).

We isolate the rest of the argument in the

LEMMA. Let A be a domain for which rank one projectives are free. Then A is a G.C.D. domain if and only if the projective dimension of each ideal generated by two elements is at most one.

REMARK. If projectives are not assumed to be free in (2.2) one still has that every finitely generated ideal I admits a decomposition I = P.J, with P projective and J overdense.

The main use of (1.2) for coherent rings of global dimension two is the following

COROLLARY 2.3. Let I be a finitely generated overdense ideal of A; then A/I is an artinian ring. In particular the prime ideals above I are finitely generated.

PROOF. By (1.2) the finitistic projective dimension of A/I is zero and thus by Bass'theorem [2] A/I is perfect. Since A/I is also coherent being a coherent A-module, by Chase's result [6] A/I is artinian.

COROLLARY 2.4.(Noetherianization) Let A be a ring of global dimension two with Spec(A) noetherian and A_P noetherian for each prime ideal P. Then A is noetherian.

First we consider

PROPOSITION 2.5. Let A be a ring of global dimension two with Max(A), the subspace of maximal ideals, noetherian. Then A is a finite product of domains; in particular A is coherent.

LEMMA. Let A be a commutative ring with Max(A) noetherian. Then each pure ideal I of A (i.e. A/I is A-flat) is generated by one idempotent.

PROOF. Recall that purity for the ideal I means that for each prime ideal P the localization I_P yields 0 or (1).

Let I ≠ 0 be pure. For each a ε I let V(a) be the closed subset of Max(A) defined by the ideal (a). Pick a such that V(a) is minimal. As (a) = aI (checked by localization) we may write a = ab with b ε I. Thus V(a) = V(b) and hence (a,1-b) = A. Write ra + s(1-b) = 1 and e = ra. Notice that e is an idempotent; we claim (e) = I. Clearly V(e) = V(a); with I = (e) ⊕ (1-e)I observe that (1-e)I is a pure ideal of the ring A(1-e). If there is an element c of (1-e)I not contained in every maximal ideal of A(1-e) (c exists if (1-e)I ≠ 0) then V(e+c) is properly contained in V(e), which is impossible.

The proof of (2.5) now follows : Clearly the number of idempotents of A is finite. Assume then A indecomposable; at each localization A is a domain (cf.[17]) and thus the minimal primes are pure ideals.

PROOF of (2.4): By the preceding we may assume A to be a domain. By Heinzer-Ohm's theorem [9] it is enough to show that for each finitely generated ideal I the set of (weakly) associated primes of A/I is finite. Write I = P.J with P projective and J overdense. Since it is enough to prove this locally in the Zariski topology, assume P = (d). Thus

$$0 \longrightarrow (d)/dJ \longrightarrow A/I \longrightarrow A/(d) \longrightarrow 0$$

yields Ass(A/I) ⊂ Ass(A/(d)) ∪ Ass(A/J). By (2.3) Ass(A/J) is finite while a prime ideal is associated to A/(d) if and only if it is minimal over (d) since A is locally noetherian.

REFERENCES

[1] M.Auslander and D.Buchsbaum, Codimension and Multiplicity, Annals Math. 68 (1958), 625-657.

[2] H.Bass, Finitistic dimension and a homological generalization of semi-primary rings, Trans.Amer.Math.Soc. 95 (1960), 466-488.

[3] N.Bourbaki, Algèbre Commutative, Hermann,Paris, 1961-65.

[4] L.Burch, On ideals of finite homological dimension in local rings, Proc.Camb.Phil.Soc. 64 (1968), 941-948.

[5] H.Cartan and S.Eilenberg, Homological Algebra, Princeton University Press, 1956.

[6] S.U.Chase, Direct product of modules, Trans. Amer.Math.Soc. 97 (1960), 457-473.

[7] J.M.Cohen, A note on homological dimension, J.Algebra 11 (1969), 483-487.

[8] L.Gruson and M.Raynaud, Critères de platitude et de projectivité, Inventines Math. 13 (1971), 1-89.

[9] W.Heinzer and J.Ohm, Locally noetherian commutative rings, Trans. Amer. Math.Soc. 158 (1971), 273-284.

[10] C.U.Jensen, Some remarks on a change of rings theorem, Math. Zeit. 106 (1968), 395-401.

[11] I.Kaplansky, Commutative Rings, Allyn and Bacon, Boston, 1970.

[12] ________, Fields and Rings, University of Chicago Press, 1969.

[13] R.E.MacRae, On an application of the Fitting invariants, J. Algebra 2 (1965), 153-169.

[14] W.V.Vasconcelos, On projective modules of finite rank, Proc. Amer.Math.Soc. 22 (1969), 430-433.

[15] ________, Annihilators of modules with a finite free resolution, Proc.Amer.Math.Soc. 29 (1971), 440-442.

[16] ________, Finiteness in projective ideals, J.Algebra (to appear).

[17] ________, The local rings of global dimension two, Proc. Amer.Math.Soc. (to appear).

Lecture Notes in Mathematics

Comprehensive leaflet on request

Vol. 111: K. H. Mayer, Relationen zwischen charakteristischen Zahlen. III, 99 Seiten. 1969. DM 16,–

Vol. 112: Colloquium on Methods of Optimization. Edited by N. N. Moiseev. IV, 293 pages. 1970. DM 18,–

Vol. 113: R. Wille, Kongruenzklassengeometrien. III, 99 Seiten. 1970. DM 16,–

Vol. 114: H. Jacquet and R. P. Langlands, Automorphic Forms on GL (2). VII, 548 pages. 1970. DM 24,–

Vol. 115: K. H. Roggenkamp and V. Huber-Dyson, Lattices over Orders I. XIX, 290 pages. 1970. DM 18,–

Vol. 116: Sèminaire Pierre Lelong (Analyse) Année 1969. IV, 195 pages. 1970. DM 16,–

Vol. 117: Y. Meyer, Nombres de Pisot, Nombres de Salem et Analyse Harmonique. 63 pages. 1970. DM 16,–

Vol. 118: Proceedings of the 15th Scandinavian Congress, Oslo 1968. Edited by K. E. Aubert and W. Ljunggren. IV, 162 pages. 1970. DM 16,–

Vol. 119: M. Raynaud, Faisceaux amples sur les schémas en groupes et les espaces homogènes. III, 219 pages. 1970. DM 16,–

Vol. 120: D. Siefkes, Büchi's Monadic Second Order Successor Arithmetic. XII, 130 Seiten. 1970. DM 16,–

Vol. 121: H. S. Bear, Lectures on Gleason Parts. III, 47 pages. 1970. DM 16,–

Vol. 122: H. Zieschang, E. Vogt und H.-D. Coldewey, Flächen und ebene diskontinuierliche Gruppen. VIII, 203 Seiten. 1970. DM 16,–

Vol. 123: A. V. Jategaonkar, Left Principal Ideal Rings. VI, 145 pages. 1970. DM 16,–

Vol. 124: Séminare de Probabilités IV. Edited by P. A. Meyer. IV, 282 pages. 1970. DM 20,–

Vol. 125: Symposium on Automatic Demonstration. V, 310 pages. 1970. DM 20,–

Vol. 126: P. Schapira, Théorie des Hyperfonctions. XI, 157 pages. 1970. DM 16,–

Vol. 127: I. Stewart, Lie Algebras. IV, 97 pages. 1970. DM 16,–

Vol. 128: M. Takesaki, Tomita's Theory of Modular Hilbert Algebras and its Applications. II, 123 pages. 1970. DM 16,–

Vol. 129: K. H. Hofmann, The Duality of Compact Semigroups and C*- Bigebras. XII, 142 pages. 1970. DM 16,–

Vol. 130: F. Lorenz, Quadratische Formen über Körpern. II, 77 Seiten. 1970. DM 16,–

Vol. 131: A Borel et al., Seminar on Algebraic Groups and Related Finite Groups. VII, 321 pages. 1970. DM 22,–

Vol. 132: Symposium on Optimization. III, 348 pages. 1970. DM 22,–

Vol. 133: F. Topsøe, Topology and Measure. XIV, 79 pages. 1970. DM 16,–

Vol. 134: L. Smith, Lectures on the Eilenberg-Moore Spectral Sequence. VII, 142 pages. 1970. DM 16,–

Vol. 135: W. Stoll, Value Distribution of Holomorphic Maps into Compact Complex Manifolds. II, 267 pages. 1970. DM 18,–

Vol. 136: M. Karoubi et al., Séminaire Heidelberg-Saarbrücken-Strasbuorg sur la K-Théorie. IV, 264 pages. 1970. DM 18,–

Vol. 137: Reports of the Midwest Category Seminar IV. Edited by S. MacLane. III, 139 pages. 1970. DM 16,–

Vol. 138: D. Foata et M. Schützenberger, Théorie Géométrique des Polynômes Eulériens. V, 94 pages. 1970. DM 16,–

Vol. 139: A. Badrikian, Séminaire sur les Fonctions Aléatoires Linéaires et les Mesures Cylindriques. VII, 221 pages. 1970. DM 18,–

Vol. 140: Lectures in Modern Analysis and Applications II. Edited by C. T. Taam. VI, 119 pages. 1970. DM 16,–

Vol. 141: G. Jameson, Ordered Linear Spaces. XV, 194 pages. 1970. DM 16,–

Vol. 142: K. W. Roggenkamp, Lattices over Orders II. V, 388 pages. 1970. DM 22,–

Vol. 143: K. W. Gruenberg, Cohomological Topics in Group Theory. XIV, 275 pages. 1970. DM 20,–

Vol. 144: Seminar on Differential Equations and Dynamical Systems, II. Edited by J. A. Yorke. VIII, 268 pages. 1970. DM 20,–

Vol. 145: E. J. Dubuc, Kan Extensions in Enriched Category Theory. XVI, 173 pages. 1970. DM 16,–

Vol. 146: A. B. Altman and S. Kleiman, Introduction to Grothendieck Duality Theory. II, 192 pages. 1970. DM 18,–

Vol. 147: D. E. Dobbs, Cech Cohomological Dimensions for Commutative Rings. VI, 176 pages. 1970. DM 16,–

Vol. 148: R. Azencott, Espaces de Poisson des Groupes Localement Compacts. IX, 141 pages. 1970. DM 16,–

Vol. 149: R. G. Swan and E. G. Evans, K-Theory of Finite Groups and Orders. IV, 237 pages. 1970. DM 20,–

Vol. 150: Heyer, Dualität lokalkompakter Gruppen. XIII, 372 Seiten. 1970. DM 20,–

Vol. 151: M. Demazure et A. Grothendieck, Schémas en Groupes I. (SGA 3). XV, 562 pages. 1970. DM 24,–

Vol. 152: M. Demazure et A. Grothendieck, Schémas en Groupes II. (SGA 3). IX, 654 pages. 1970. DM 24,–

Vol. 153: M. Demazure et A. Grothendieck, Schémas en Groupes III. (SGA 3). VIII, 529 pages. 1970. DM 24,–

Vol. 154: A. Lascoux et M. Berger, Variétés Kähleriennes Compactes. VII, 83 pages. 1970. DM 16,–

Vol. 155: Several Complex Variables I, Maryland 1970. Edited by J. Horváth. IV, 214 pages. 1970. DM 18,–

Vol. 156: R. Hartshorne, Ample Subvarieties of Algebraic Varieties. XIV, 256 pages. 1970. DM 20,–

Vol. 157: T. tom Dieck, K. H. Kamps und D. Puppe, Homotopietheorie. VI, 265 Seiten. 1970. DM 20,–

Vol. 158: T. G. Ostrom, Finite Translation Planes. IV. 112 pages. 1970. DM 16,–

Vol. 159: R. Ansorge und R. Hass. Konvergenz von Differenzenverfahren für lineare und nichtlineare Anfangswertaufgaben. VIII, 145 Seiten. 1970. DM 16,–

Vol. 160: L. Sucheston, Constributions to Ergodic Theory and Probability. VII, 277 pages. 1970. DM 20,–

Vol. 161: J. Stasheff, H-Spaces from a Homotopy Point of View. VI, 95 pages. 1970. DM 16,–

Vol. 162: Harish-Chandra and van Dijk, Harmonic Analysis on Reductive p-adic Groups. IV, 125 pages. 1970. DM 16,–

Vol. 163: P. Deligne, Equations Différentielles à Points Singuliers Reguliers. III, 133 pages. 1970. DM 16,–

Vol. 164: J. P. Ferrier, Seminaire sur les Algebres Complètes. II, 69 pages. 1970. DM 16,–

Vol. 165: J. M. Cohen, Stable Homotopy. V, 194 pages. 1970. DM 16,–

Vol. 166: A. J. Silberger, PGL_2 over the p-adics: its Representations, Spherical Functions, and Fourier Analysis. VII, 202 pages. 1970. DM 18,–

Vol. 167: Lavrentiev, Romanov and Vasiliev, Multidimensional Inverse Problems for Differential Equations. V, 59 pages. 1970. DM 16,–

Vol. 168: F. P. Peterson, The Steenrod Algebra and its Applications: A conference to Celebrate N. E. Steenrod's Sixtieth Birthday. VII, 317 pages. 1970. DM 22,–

Vol. 169: M. Raynaud, Anneaux Locaux Henséliens. V, 129 pages. 1970. DM 16,–

Vol. 170: Lectures in Modern Analysis and Applications III. Edited by C. T. Taam. VI, 213 pages. 1970. DM 18,–

Vol. 171: Set-Valued Mappings, Selections and Topological Properties of 2^X. Edited by W. M. Fleischman. X, 110 pages. 1970. DM 16,–

Vol. 172: Y.-T. Siu and G. Trautmann, Gap-Sheaves and Extension of Coherent Analytic Subsheaves. V, 172 pages. 1971. DM 16,–

Vol. 173: J. N. Mordeson and B. Vinograde, Structure of Arbitrary Purely Inseparable Extension Fields. IV, 138 pages. 1970. DM 16,–

Vol. 174: B. Iversen, Linear Determinants with Applications to the Picard Scheme of a Family of Algebraic Curves. VI, 69 pages. 1970. DM 16,–

Vol. 175: M. Brelot, On Topologies and Boundaries in Potential Theory. VI, 176 pages. 1971. DM 18,–

Vol. 176: H. Popp, Fundamentalgruppen algebraischer Mannigfaltigkeiten. IV, 154 Seiten. 1970. DM 16,–

Vol. 177: J. Lambek, Torsion Theories, Additive Semantics and Rings of Quotients. VI, 94 pages. 1971. DM 16,–

Please turn over

Vol. 178: Th. Bröcker und T. tom Dieck, Kobordismentheorie. XVI, 191 Seiten. 1970. DM 18,–

Vol. 179: Seminaire Bourbaki – vol. 1968/69. Exposés 347-363. IV. 295 pages. 1971. DM 22,–

Vol. 180: Séminaire Bourbaki – vol. 1969/70. Exposés 364-381. IV, 310 pages. 1971. DM 22,–

Vol. 181: F. DeMeyer and E. Ingraham, Separable Algebras over Commutative Rings. V, 157 pages. 1971. DM 16.–

Vol. 182: L. D. Baumert. Cyclic Difference Sets. VI, 166 pages. 1971. DM 16,–

Vol. 183: Analytic Theory of Differential Equations. Edited by P. F. Hsieh and A. W. J. Stoddart. VI, 225 pages. 1971. DM 20,–

Vol. 184: Symposium on Several Complex Variables, Park City, Utah, 1970. Edited by R. M. Brooks. V, 234 pages. 1971. DM 20,–

Vol. 185: Several Complex Variables II, Maryland 1970. Edited by J. Horváth. III, 287 pages. 1971. DM 24.–

Vol. 186: Recent Trends in Graph Theory. Edited by M. Capobianco/ J. B. Frechen/M. Krolik. VI, 219 pages. 1971. DM 18.–

Vol. 187: H. S. Shapiro, Topics in Approximation Theory. VIII, 275 pages. 1971. DM 22,–

Vol. 188: Symposium on Semantics of Algorithmic Languages. Edited by E. Engeler. VI, 372 pages. 1971. DM 26,–

Vol. 189: A. Weil, Dirichlet Series and Automorphic Forms. V. 164 pages. 1971. DM 16,–

Vol. 190: Martingales. A Report on a Meeting at Oberwolfach, May 17-23, 1970. Edited by H. Dinges. V, 75 pages. 1971. DM 16,–

Vol. 191: Séminaire de Probabilités V. Edited by P. A. Meyer. IV, 372 pages. 1971. DM 26,–

Vol. 192: Proceedings of Liverpool Singularities – Symposium I. Edited by C. T. C. Wall. V, 319 pages. 1971. DM 24,–

Vol. 193: Symposium on the Theory of Numerical Analysis. Edited by J. Ll. Morris. VI, 152 pages. 1971. DM 16,–

Vol. 194: M. Berger, P. Gauduchon et E. Mazet. Le Spectre d'une Variété Riemannienne. VII, 251 pages. 1971. DM 22,–

Vol. 195: Reports of the Midwest Category Seminar V. Edited by J.W. Gray and S. Mac Lane.III, 255 pages. 1971. DM 22,–

Vol. 196: H-spaces – Neuchâtel (Suisse)- Août 1970. Edited by F. Sigrist, V, 156 pages. 1971. DM 16,–

Vol. 197: Manifolds – Amsterdam 1970. Edited by N. H. Kuiper. V, 231 pages. 1971. DM 20,–

Vol. 198: M. Hervé, Analytic and Plurisubharmonic Functions in Finite and Infinite Dimensional Spaces. VI, 90 pages. 1971. DM 16.–

Vol. 199: Ch. J. Mozzochi, On the Pointwise Convergence of Fourier Series. VII, 87 pages. 1971. DM 16,–

Vol. 200: U. Neri, Singular Integrals. VII, 272 pages. 1971. DM 22,–

Vol. 201: J. H. van Lint, Coding Theory. VII, 136 pages. 1971. DM 16,–

Vol. 202: J. Benedetto, Harmonic Analysis on Totally Disconnected Sets. VIII, 261 pages. 1971. DM 22,–

Vol. 203: D. Knutson, Algebraic Spaces. VI, 261 pages. 1971. DM 22,–

Vol. 204: A. Zygmund, Intégrales Singulières. IV, 53 pages. 1971. DM 16,–

Vol. 205: Séminaire Pierre Lelong (Analyse) Année 1970. VI, 243 pages. 1971. DM 20,–

Vol. 206: Symposium on Differential Equations and Dynamical Systems. Edited by D. Chillingworth. XI, 173 pages. 1971. DM 16,–

Vol. 207: L. Bernstein, The Jacobi-Perron Algorithm – Its Theory and Application. IV, 161 pages. 1971. DM 16,–

Vol. 208: A. Grothendieck and J. P. Murre, The Tame Fundamental Group of a Formal Neighbourhood of a Divisor with Normal Crossings on a Scheme. VIII, 133 pages. 1971. DM 16,–

Vol. 209: Proceedings of Liverpool Singularities Symposium II. Edited by C. T. C. Wall. V, 280 pages. 1971. DM 22,–

Vol. 210: M. Eichler, Projective Varieties and Modular Forms. III, 118 pages. 1971. DM 16,–

Vol. 211: Théorie des Matroïdes. Edité par C. P. Bruter. III, 108 pages. 1971. DM 16,–

Vol. 212: B. Scarpellini, Proof Theory and Intuitionistic Systems. VII, 291 pages. 1971. DM 24,–

Vol. 213: H. Hogbe-Nlend, Théorie des Bornologies et Applications. V, 168 pages. 1971. DM 18,–

Vol. 214: M. Smorodinsky, Ergodic Theory, Entropy. V, 64 pages. 1971. DM 16,–

Vol. 215: P. Antonelli, D. Burghelea and P. J. Kahn, The Concordance-Homotopy Groups of Geometric Automorphism Groups. X, 140 pages 1971. DM 16,–

Vol. 216: H. Maaß, Siegel's Modular Forms and Dirichlet Series. VII, 328 pages. 1971. DM 20,–

Vol. 217: T. J. Jech, Lectures in Set Theory with Particular Emphasis on the Method of Forcing. V, 137 pages. 1971. DM 16,–

Vol. 218: C. P. Schnorr, Zufälligkeit und Wahrscheinlichkeit. IV, 212 Seiten 1971. DM 20,–

Vol. 219: N. L. Alling and N. Greenleaf, Foundations of the Theory of Klein Surfaces. IX, 117 pages. 1971. DM 16,–

Vol. 220: W. A. Coppel, Disconjugacy. V, 148 pages. 1971. DM 16,–

Vol. 221: P. Gabriel und F. Ulmer, Lokal präsentierbare Kategorien. V, 200 Seiten. 1971. DM 18,–

Vol. 222: C. Meghea, Compactification des Espaces Harmoniques. III, 108 pages. 1971. DM 16,–

Vol. 223: U. Felgner, Models of ZF-Set Theory. VI, 173 pages. 1971. DM 16,–

Vol. 224: Revètements Etales et Groupe Fondamental. (SGA 1). Dirigé par A. Grothendieck XXII, 447 pages. 1971. DM 30,–

Vol. 225: Théorie des Intersections et Théorème de Riemann-Roch. (SGA 6). Dirigé par P. Berthelot, A. Grothendieck et L. Illusie. XII, 700 pages. 1971. DM 40,–

Vol. 226: Seminar on Potential Theory, II. Edited by H. Bauer. IV, 170 pages. 1971. DM 18,–

Vol. 227: H. L. Montgomery, Topics in Multiplicative Number Theory. IX, 178 pages. 1971. DM 18,–

Vol. 228: Conference on Applications of Numerical Analysis. Edited by J. Ll. Morris. X, 358 pages. 1971. DM 26,–

Vol. 229: J. Väisälä, Lectures on n-Dimensional Quasiconforma Mappings. XIV, 144 pages. 1971. DM 16,–

Vol. 230: L. Waelbroeck, Topological Vector Spaces and Algebras. VII, 158 pages. 1971. DM 16,–

Vol. 231: H. Reiter, L^1-Algebras and Segal Algebras. XI, 113 pages 1971. DM 16,–

Vol. 232: T. H. Ganelius, Tauberian Remainder Theorems. VI, 75 pages. 1971. DM 16,–

Vol. 233: C. P. Tsokos and W. J. Padgett. Random Integral Equations with Applications to Stochastic Systems. VII, 174 pages. 1971. DM 18,–

Vol. 234: A. Andreotti and W. Stoll. Analytic and Algebraic Dependence of Meromorphic Functions. III, 390 pages. 1971. DM 26,–

Vol. 235: Global Differentiable Dynamics. Edited by O. Hájek, A. J. Lohwater, and R. McCann. X, 140 pages. 1971. DM 16,–

Vol. 236: M. Barr, P. A. Grillet, and D. H. van Osdol. Exact Categories and Categories of Sheaves. VII, 239 pages. 1971. DM 20,–

Vol. 237: B. Stenström. Rings and Modules of Quotients. VII, 136 pages. 1971. DM 16,–

Vol. 238: Der kanonische Modul eines Cohen-Macaulay-Rings. Herausgegeben von Jürgen Herzog und Ernst Kunz. VI, 103 Seiten 1971. DM 16,–

Vol. 239: L. Illusie, Complexe Cotangent et Déformations I. XV, 355 pages. 1971. DM 26,–

Vol. 240: A. Kerber, Representations of Permutation Groups I. VII 192 pages. 1971. DM 18,–

Vol. 241: S. Kaneyuki, Homogeneous Bounded Domains and Siege Domains. V, 89 pages. 1971. DM 16,–

Vol. 242: R. R. Coifman et G. Weiss, Analyse Harmonique Non Commutative sur Certains Espaces. V, 160 pages. 1971. DM 16,–

Vol. 243: Japan-United States Seminar on Ordinary Differential and Functional Equations. Edited by M. Urabe. VIII, 332 pages. 1971 DM 26,–

Vol. 244: Séminaire Bourbaki – vol. 1970/71. Exposés 382-399 IV, 356 pages. 1971. DM 26,–

Vol. 245: D. E. Cohen, Groups of Cohomological Dimension One. V 99 pages. 1972. DM 16,–